D0850703

Galileo's Glassworks

Galileo's Glassworks

The Telescope and the Mirror

Eileen Reeves

Harvard University Press

Cambridge, Massachusetts

London, England

2008

Library of Congress Cataloging-in-Publication Data

Reeves, Eileen Adair.
Galileo's glassworks : the telescope and the mirror /
Eileen Reeves.
p.   cm.
Includes bibliographical references and index.
ISBN-13: 978-0-674-02667-4 (alk. paper)
ISBN-10: 0-674-02667-5 (alk. paper)
1. Astronomical instruments—Europe—History—17th century.
2. Telescopes—Europe—History—17th century.
3. Optical instruments—Europe—History—17th century.
4. Mirrors—Experiments—History—17th century.
5. Galilei, Galileo, 1564–1642.
6. Science—History—17th century.
I. Title.
QB85.8.R44 2008
522′.2092—dc22   2007005089

Contents

———❖———

Galileo's Glassworks

The Hague, 1608

$\mathcal{T}$HE Dutch telescope and the Italian scientist Galileo Galilei have enjoyed a durable connection in the popular mind, so much so that one might argue that it was this simple instrument that transformed a rather modest middle-aged scholar and tutor in Padua into Europe's best-known private citizen, the bold icon of the Copernican Revolution, and the most celebrated casualty of Counter-Reformation science. The telescope appears to have changed Galileo's life and the course of early modern astronomy with extraordinary rapidity: about eighteen months elapsed between the invention of the instrument in The Hague and the publication of Galileo's *Starry Messenger* in Venice, and less than two years passed before he left Padua for Florence to become Mathematician and Philosopher at the Court of the Grand Duke of Tuscany.[1] The velocity and magnitude of these events, however, mask the astronomer's own tardy and curiously obscured encounter with the Dutch instrument. The record suggests that Galileo, like several of his peers, initially misunderstood the basic design of the telescope. This book is concerned, therefore, with two fundamental questions of intellectual accountability: What did Galileo know of the invention, and when did he know it?

In seeking to explain the gap between the telescope's emergence in the Dutch Republic in late September 1608—news of which had reached Galileo's close associate Paolo Sarpi in Venice within six or seven weeks—and the astronomer's alleged acquaintance with it sometime between mid-May and late July of 1609, this book considers how and why information about the telescope was transmitted, suppressed, garbled, or misconstrued en route. Historians of science and biographers have discreetly passed over these seemingly unproductive and poorly documented months, expressing at most mild amazement at Sarpi's apparent failure to relay the rumor, at others' timelier encounter with the device elsewhere in Italy, or at the confused chronology provided by Galileo then and in later years.[2] I am offering a revised version of events, one that rejects these unlikely narratives of silence and idleness in favor of an account privileging the role of misinterpretation, error, and preconception.

News of a Novelty

News of the telescope—an excited insistence on the novelty of the device, but only the barest allusion to its optical components—reached the authorities in the Dutch Republic and a wider European public through several channels. On September 25, 1608, the Committee of Councilors of the Province of Zeeland issued a memo stating that "the bearer" claimed to have invented an instrument for seeing remote objects as if they were near, and that he requested the opportunity to demonstrate it to Prince Maurice of Nassau.[3] This unnamed bearer is entirely typical of the narrative: references to *quīdam Belga* or "a certain Netherlander" found their place in many early discussions of the invention and suggested to some readers in that bellicose age that the device had emerged, not in the Dutch Republic, but amid its bitterest enemies, in the Spanish Netherlands.[4]

On October 2, 1608, just one week after this memo, a spectacle maker residing in the Dutch city of Middelburg, Hans Lipperhey, submitted a patent application for binocular telescopes with quartz rather than mere glass lenses. His negotiations, though remunerative, were ultimately unsuccessful, for within two more weeks the Councilors of Zeeland were negotiating with another individual who claimed, not to have invented the telescope, but merely to know how to manufacture it. The following day, a third contender, Jacob Metius of Alkmaar, petitioned the authorities in writing for an exclusive patent for a telescopic device. While acknowledging the recent efforts of a spectacle-maker of Middelburg—presumably Hans Lipperhey—Metius objected that he himself had a prior claim, for he had been conducting his own research for two years and was privy to secret knowledge about glass once available to "some of the ancients." Two days later, on October 17, he was granted a sum to improve his prototype and was invited to apply for a patent once it had been brought to perfection.[5]

Even as word of such developments was spreading to princely courts and into print, a dealer was already offering his telescope for sale. This individual, ambiguously identified as "the original Netherlandish inventor" of the device, sought to sell an instrument with a cracked lens at the Frankfurt Fair sometime in the fall of 1608, quite possibly *before* the invention emerged in The Hague. At least one potential buyer examined it, but he demurred, finally, in the face of a price he judged exorbitant. The would-be consumer naturally began to try to replicate the device upon his return to Bavaria.[6]

Because Prince Maurice had examined the telescope in The Hague in the company of his enemy, Marquis Ambrogio Spinola, commander in chief of the military forces of the Spanish Netherlands, the latter soon demonstrated the invention to the ruler of the Spanish Netherlands, Archduke Albert of Austria, who appears to have obtained two more that winter and spring. The papal nuncio Guido Bentivoglio, a former student of Galileo's, used the archduke's telescope in this period, described the experience in a letter to Cardinal Scipione Borghese, nephew of Pope Paul V, and is the likely source of the instrument sent to the cardinal in Rome.[7] And the fact that the English ambassador to the Spanish Netherlands suddenly took up astronomy in mid-April 1609 on the occasion of his visits to the archduke's castle suggests that he, too, now had access to the new invention.[8]

A French-language newsletter originating in The Hague in early October 1608 and devoted, for the most part, to an exotic description of the embassy of the King of Siam to the Dutch Republic, also included a brief article about "certain glasses" allowing clear vision of features of landscapes five or even ten miles away. The article attributed the invention to a "poor, pious and God-fearing" lens maker in Middelburg, and noted ominously that the device had been made available both to Prince Maurice and to his enemy General Spinola.[9] By late November 1608 the newsletter had reached Venice, and Paolo Sarpi, whose correspondence minimizes its importance, emphasizes the meager results of his own prior research with lens-and-mirror combinations and makes no mention of discussing the invention with Galileo in this period.

It is clear, however, that during the period from late autumn 1608 to the spring of 1609 the telescope was becoming increasingly well known far beyond its point of origin. The article describing its emergence in the Dutch Republic and its instant availability to the Spanish Netherlands was reprinted in Lyons in November 1608, and in December the French diplomat Pierre Jeannin sent a soldier from The Hague to Henri IV and the Duke de Sully in order that he might manufacture two such instruments for the king and his minister. Henri, for his part, appears to have regarded the instrument as a distracting trinket, saying that he needed at present to focus on nearby rather than remote matters.[10] But by April 1609 the telescope was commercially available in Paris in a shop on the Pont Neuf, by May it was in the possession of the Count of Fuentes, the Spanish governor of Milan, and by sometime that summer it appeared in cities as distant as Ansbach in Bavaria, London, Rome, Naples, Venice, and Padua.[11]

It was not until late that spring or summer, Galileo would write in his *Starry Messenger*, that he first heard the rumor of the telescope, seeing confirmation of the news in an undated letter from Jacques Badovere, a former student in Paris, a day or two later, and setting to work on the device immediately thereafter. In a second explanation dating to 1623, Galileo referred again to his belated acquaintance with the bare rumor of the telescope's existence, but discarded the detail about the obliging middleman in Paris, and added that he successfully made the device within a few days of hearing the now rather stale news from The Hague.

The story of the telescope's emergence in The Hague and of the rapid diffusion of both the device and the rumor gives rise to two related questions. First, and most obviously, why did Galileo appear to remain unacquainted with such news for so long? Secondly, and less obviously, why do allusions to the telescope's relative antiquity begin to surface amid narratives that stress, for the most part, the instrument's novelty and the priority claims of its various inventors? The suggestion that the telescope was *not* an entirely new invention in the fall of 1608 gained momentum over the coming months and years, and as this book will show, the widespread impression of the telescope's senescence is directly related to the curious gap we find in Galileo's own versions of the narrative.

Déjà Vu

Some of those concerned with the emergence of the telescope implied that a similar instrument, now lost, had been the possession of the ancients. Jacob Metius, for instance, explicitly associated his invention with scrutiny of the more arcane uses of glass in antiquity, and the *Mercure François* of 1611, in commenting upon the events at The Hague, recalled the medieval philosopher

Roger Bacon's statement that Julius Caesar had used mirrors in Gaul to view his enemies in England, and noted with regret that many other "beautiful inventions had been lost."[12]

Other writers, though making no claim for the telescope's antiquity, insisted on backdating its invention a generation or two. The most striking such instance was offered by Girolamo Sirtori, the man who examined a telescope upon its arrival in Milan in May 1609: the youthful and vigorous protagonists of his account, the putative inventors and diffusers of new optical expertise, undergo perpetual displacement by a band of older forerunners. Here, for example, Hans Lipperhey comes by his knowledge simply by observing the way in which a more informed customer manipulates a pair of convex and concave lenses in his shop, and when the spectacle maker races with an invention that is clearly *not* his to Prince Maurice, the Dutch ruler in turn appears to have hidden his own prior familiarity with and possible possession of such a device. Several months later, in this animated account, a Frenchman rushes to Milan with the instrument, and Sirtori, eager to discover the theoretical basis of the telescope, likewise undertakes a series of headlong adventures, squandering money on an imperfect lens in Venice, and imprudently trying out the device in the Tower of Saint Mark, eventually finding himself surrounded for hours by a gang of noble youths eager to use it as well. Traveling to Spain, Sirtori encounters ever more extravagant rumors of the telescope, and in time he meets a withered old man who reveals a set of rusted lens-making tools and a manuscript in a monkish hand describing in detail the whole lost art of optics, and in particular the crucial curvatures of the lenses. Because the old man allows him to copy just enough information to copy these curvatures, Sirtori has what functions (albeit only on paper) as a successful forerunner of the Dutch telescope. When in 1611 the powerful Archduke Maximilian of Bavaria shows him a similar illustration, made from a telescope lately sent by Galileo, Sirtori produces this drawing, with its aura of antiquity and priority, by way of response.[13]

Other allusions to the prehistory of the Dutch telescope emerged within the context of Galileo's success and appear to be at least partially motivated by a desire to diminish or share in the astronomer's fame. The Florentine poet Raffaello Gualterotti, for example, claimed in the spring of 1610, within weeks of the publication of Galileo's *Starry Messenger,* that twelve years earlier he had made and neglected "a puny thing" offering telescopic vision, insisting that it was a chiefly matter of regional pride that made him begrudge the credit given to "some Fleming." And in the summer of 1609 and for the rest of his life, the elderly Neapolitan playwright and natural philosopher Giambattista della Porta would gesture with scorn to the "secret of the telescope," alleging that the pertinent optical details came from a technical work on refraction he had published in 1593.[14]

Those who did not advance their own candidacy as inventors often alluded, in fact, to *Natural Magic,* a widely read and less technical work published by della Porta in 1589. In the spring of 1610, for instance, a Frenchman reported to Paolo Sarpi that a myopic friend with no particular skill in optics or astronomy was able to see lunar craters with a telescope he had just constructed from clues found in della Porta's popular and technical works.

Around the same time, and likewise in response to Galileo's publication, Johannes Kepler noted in print that the recent Dutch invention had been "made public" years earlier in *Natural Magic*.[15] In the coming months Kepler's observation was both attacked and endorsed by many writers, one of whom seemed to convert the matter into an obscure referendum on the grasping Galileo and a man "honored by years and white hair."[16] Over a decade later, della Porta's role in the invention still proved grounds for a fine argument between an Italian and an anonymous Dutchman who claimed the device as his own.[17] And as late as 1641 the French scholar Pierre Gassendi would maintain that although the actual invention had been made "by chance" by Metius, the idea for a similar one had been publicized years earlier by della Porta.[18]

What is remarkable about these and other references to forerunners of the Dutch telescope is that they cannot be neatly divided into improbable allusions to a telescopic device in antiquity and more feasible gestures to instruments dating from the mid- to late sixteenth century. To be sure, the latter group of claims is certainly more detailed: Sirtori named his elderly Spanish informant, for instance, and della Porta and his followers often mentioned specific passages in his works in support of their claims, generally pointing to his discussion of lenses and occasionally to his treatment of mirrors as well. But in his *Natural Magic* of 1589 della Porta himself had insisted upon the prior existence of a telescopic instrument in antiquity, the celebrated Pharos or lighthouse of Alexandria, and used it as the basis of his own discussion of lens-and-mirror combinations. The first detailed description of the Dutch telescope, published in Latin in the fall of 1609, offered information about the two glass lenses and tube crucial to the device, but also rehearsed within this context, as if fundamentally inseparable from the recent developments in The Hague, the legendary claims concerning the Pharos drawn from della Porta's work.[19] And even so sober a writer as the Sienese engineer Sergio Venturi would maintain in early 1610 both that della Porta's technical work of 1593 contained the theoretical basis of the telescope, *and* that the device had been known to the ancients, who had kept such knowledge out of "the hands of the common man."[20] As this book will show, popular beliefs about telescopic implements of antiquity both informed and deformed early modern optical research, such that they cannot be easily dissociated from claims about the Dutch telescope and its alleged forerunners. This conflation of the fabulous remote past and the plausible recent past underwrote many of the earliest impressions of the telescope, among them Galileo's.

The Imperial Mirror

The early Dutch telescope, a twelve- or fourteen-inch tube enclosing a convex glass objective and a concave glass eyepiece and initially magnifying remote objects only about three times, benefited from two widespread traditions concerning enhanced vision, for these provided a ready-made vernacular for the new and often quite disappointing device. First of all, the telescope was described as a superior version of common reading glasses, as if it had been designed above all for the remote and potentially covert processing of textual information, or more generally for espionage. It is thus not surprising to find what would eventually be called the "spyglass" in English explicitly identified as an *espía de vidrio,* a "glass spy," in a polemical Spanish work written in the 1630s.[21] Less predictably, the Dutch telescope was also compared to a legendary mirror or combination of lenses and mirrors with which ancient rulers had supposedly surveyed foreign enemies and their own fractious empires. Though both traditions had a certain patina of the antique—the magnification of letters seen through a water-filled container having been noted in the first century A.D. by Seneca, and the "imperial mirror" enjoying an occasional connection to Virgil, Alexander the Great, Apollonius, Ptolemy, Archimedes, and Hercules—they arose alike in the literature of the medieval period, and gained new currency in the context of optical research undertaken in the decades just before and after the Dutch invention.[22]

It is the latter legend, that of the telescopic mirror, that warrants our closer attention here, for Galileo's misapprehension of the news from The Hague likely involves his belief that the device incorporated a mirror. There were two lens-and-mirror combinations under scrutiny in the early modern period. One was strongly associated with recent innovations upon the *camera obscura* and, like the configuration in the dark room, involved a convex glass lens trained upon an object and a concave mirror that was set at an acute angle to the lens and served as an eyepiece. This combination was somewhat unwieldy—for the observer cannot face the object or the lens, but only the eyepiece,

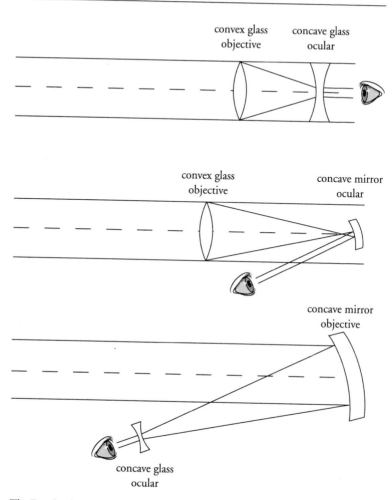

The Dutch telescope and two lens-and-mirror combinations. Line drawing by Margaret Nelson.

and his or her head, unless tilted, will impede the image emitted by the lens—and would have yielded only slightly magnified images and a relatively small field of view.

The other lens-and-mirror combination appears to have emerged in the context of surveying. Surveyors sometimes used plane mirrors or even basins of water in which large and distant objects such as towers were reflected in order to calculate height. Drawing on the rectilinear nature of light rays, and on the equality of the angles of incidence and reflection, the surveyor established ratios between the large right triangle formed by the tower, the ground, and the mirror, and the smaller right triangle bounded by his own upright body, the mirror, and the ground. The basic formula had many variants, and occasionally the mirror was placed vertically. In one such arrangement, the observer substituted a concave mirror, which produced an enlarged but blurry image of the object under scrutiny. Either a convex glass lens placed slightly behind the focal point of the mirror or a concave glass lens slightly in front of it would have sharpened this magnified image. The latter arrangement, which the observer would be able to adjust by moving the lens away from his eye and toward the mirror, would have been somewhat easier to achieve.[23] However, the rough symmetry of the former arrangement with the second configuration—in essence, an exchange of objective and eyepiece—might have made it the more intuitive combination.

Both sorts of lens-and-mirror combinations were described, if not constructed, by optical theorists in mid- to late-sixteenth-century Italy and England, and they would have clearly differed in design and in some results from the Dutch telescope. As this book will show, the three devices enjoyed a certain conflation in the early modern period, offering the iconic instrument of the Scientific Revolution an ancient, temporary, and wholly unmerited pedigree, and the modest lens-and-mirror devices a brief afterlife in the company of their more successful replacement.

At issue, then, are not merely the brevity and vagueness of the information from The Hague, but also the variety of clues that disposed Galileo, among others, to misunderstand the news of the invention. Several centuries of literary discussions about telescopic vision, and the tardy influx of travelers' talk about ancient Alexandrian mirrors with such capabilities, formed an essential backdrop both to sixteenth-century optical experimentation and to the original expectations and misinterpretations of the Dutch device. To put it differently, the first responses to the decidedly genuine telescope in 1608 and 1609 and beyond cannot be dissociated from the wealth of cultural material about the wholly fictional imperial mirror.

Chapter One

———⟫•◈•⟪———

The Daily Mirror of Empire

*T*HE notion of telescopic vision long predates the telescope. The idea of a telescopic mirror emerged in literary and philosophical works of the medieval period, and there was an ongoing discussion of such mirrors in the early modern era. Many allusions to such mirrors are characterized by nostalgia: the wondrous devices of antiquity are always busted, rusted, or in some way defunct, and their disappearance often coincides with the end of an overstretched empire. Increasingly, however, writers began to suggest that these instruments were not the props of an irretrievable imperial past, but rather that various medieval figures—most typically, the Franciscan Roger Bacon—had deployed them, or that these devices were still working in regions remote from the reader, or, finally, that they were currently available to the more astute members of a European audience. This importation of the telescopic mirror into the present corresponds, but only rather roughly, to actual developments in natural philosophy and in the manufacture of glass and steel.

Imperial Mirrors

A tale popular throughout thirteenth- and fourteenth-century European literature insists upon the existence and eventual destruction of a telescopic mirror fashioned by Virgil, or occasionally by Merlin, and designed to protect Rome from her foreign enemies.[1] Although stories of this sort draw on the poet's odd (and wholly fictive) sideline as a magician, the fact that Virgil was strongly associated with Augustus, Rome's first emperor, adds a certain logic to the attribution. The mirror was mounted on top of a tall tower in Rome, such that "all those who looked at it / From a day's travel away could see / Every human creature / That desired or attempted / To hurt or harm Rome."[2]

The idea was evidently familiar enough to some writers to be treated in passing, or modified for political purposes, or offered as a point of departure for other and more exotic mirrors. In *The Destruction of Rome,* for example, the looking glass "of which men have said so much," appears to function simply because of its elevated position, whereas in the rhymed Italian version of *The Romance of the Seven Sages* the real enemies are Rome's rebellious provinces, and in his work the fourteenth-century French poet Jean Froissart compares himself to "the master / Who made the mirror at Rome" but affirms that were such a device his, it would be used to spy out his beloved. The eventual destruction of the Virgilian mirror is typically brought about by foreign subterfuge: the Romans, distinguished by a combination of greediness and naïveté, are at last convinced by outsiders to dig for treasure beneath the tower, and they accidentally topple the structure on which their security depends. The fourteenth-century chronicler Jean d'Outremeuse offered the conventional observation that "if the Romans had guarded this mirror well, they would still be the rulers of the world."[3]

A tardy variant on this motif associated the device with Hercules, rather than Virgil, and with the Spanish port city of La Coruña, rather than Rome.[4] The Galician tower on which the mirror was placed had much earlier been portrayed by Paulus Orosius as one with remarkable defensive possibilities, for it had been built *ad speculum Britanniae,* "for watching Britain."[5] According to the influential fourteenth-century chronicler Leomarte in his *Sumas de Hystoria Troyana,* and subsequently Raoul Lefèvre in his *Recueil des Histoires de Troye* of 1464, after founding La Coruña, Hercules had adorned the tower with a lamp that burned day and night for three hundred years, and a copper statue looking out to sea and bearing an enchanted mirror for protection against invasion.[6] The defense system worked well until the time of Nebuchadnezzar, when that ambitious Babylonian ruler, or alternatively Chaldeans fleeing him and advancing upon La Coruña, camouflaged ships with branches and green boughs. The inhabitants of the city, oddly unperturbed, took the approaching fleet to be a floating mountain; upon his arrival, the invader destroyed both mirror and lamp. The peculiar story, somewhat comic in its provinciality, became especially familiar in Britain, for William Caxton's translation of the *Recueil,* completed in 1471 and printed in Bruges around 1475, is to all appearances the first work published in English, and went through numerous re-editions for the next three centuries.[7]

The imperial mirror was not, however, said to be the sole possession of the West. The *Letter of Prester John,* which had appeared about 1165 and described the possessions of a fabulously wealthy and entirely fabricated Christian ruler of Asia, also mentions a very similar means of protecting the state. On the top of the highest column of a palace, the letter relates, "is a mirror, made in such a way, that all machinations, and everything that is done for us and against us in adjacent provinces and in those subjected to us, can be very clearly seen and known by watchmen."[8] The claims made in Latin in the very popular *Letter* were repeated and elaborated in manuscript form in Anglo-Norman, Old French, Provençal, Italian, Irish, Hebrew, German, Russian, and Serbian, and eventually printed in Venice in 1478. A detail shared by virtually all versions is the claim that one reached Prester John's mirror in an "immense and beautiful tower" by climbing 125 steps made of precious stones and confronting a vast body of armed men who ensured the safety of the optical instrument.[9] The mirror itself, in one fourteenth-century German version, was of the clearest crystal, worked "by day and by night, early and late," and showed objects both near and far; some redactions affirmed that men a week's travel from the city could perceive it.[10] Prester John consulted it each morning to learn of and to correct possible threats to his kingdom. The Old French and Occitan versions, for instance, reported that

> There is no land so distant
> Where someone desires to make war on us
> Nor treason by any people

That we do not see immediately.
We have no need of spies
Always to be giving us news
Because we see everything in the mirror,
Our enemies and their preparations,
Nothing can be hidden from us.[11]

Around 1210 the German poet Wolfram von Eschenbach alluded rather pointedly to the similarity of the Eastern and Virgilian motifs in his *Parzival,* which features a polished column with all the telescopic functions of both mirrors, and an extraordinarily elaborate architectural setting like that of Prester John's.[12] Significantly, the device is at the castle of a magician whose forefather was Virgil, but it had been stolen from an Eastern queen, and toward the end of his romance the poet alleges that Parzival's Eastern half-brother Feirefiz is to become the father of Prester John. Wolfram's reference to the outright appropriation of the literary motif is perhaps related to his relatively unusual insistence on the wholly magical aspect of the mirror; put differently, the fact that this object could be transferred, as needed, from Eastern to Western settings, suggested to this particular poet that it was no mere physical artifact, and not governed by optical principles alone. It is irreproducible in the sense that it can only be relocated, not duplicated, and Wolfram explicitly observes of the castle in which the polished column is the most marvelous feature that "the skill that went to make it would have surpassed the understanding of Master Geometras, had he set his hand to it; for it was contrived of subtle arts."[13] Whoever this Master Geometer was—and names of this sort were applied to Euclid, Archimedes, and Apollonius of Perga, all of whom contributed in antiquity to the study of catoptrics—he had nothing to do with the construction of Wolfram's device.[14]

The mirror described by Geoffrey Chaucer in "The Squire's Tale," an orientalizing romance or perhaps a parody of such, also draws upon both the Virgilian and pseudo-Oriental sources, but the poet's treatment of the problem of multiple sources for rather singular objects is quite different.[15] A knight in service to "the kyng of Arabe and of Inde" arrives in the "land of Tartarye"—the Mongol empire—and seeks out its ruler, the fabulously wealthy and wise Cambyuskan, or Genghis Khan. He bears, among other gifts, a "broad mirror of glass" and claims that men may see in it any evil that befalls them or their reign and discern with ease their friends and foes.[16] As is the case in the work of his contemporary Jean Froissart, and in the probable interest of the amorous plot of the unfinished "Squire's Tale," the mirror's use is diversified, for ladies may employ it to detect treasonous lovers. An unlettered populace greeted the knight's gifts with mistrust, some fearing the use of magic,[17] but others, Chaucer noted:

> wondred on the mirour
> That born was up into the maister tour,
> Hou men myghte in it swiche thynges se.
> Another answerde and seyde it myghte wel be
> Naturelly, by composiciouns
> Of anglis and of slye reflexiouns,
> And seyden that in Rome was swich oon.
> They speken of Alocen, and Vitulon,
> And Aristotle, that writen in hir lyves

> Of queynte mirours and of perspectives,
> As knowen they that han hir bookes herd.[18]

Some of this optical shoptalk had already been rehearsed in the thirteenth-century *Romance of the Rose,* which Chaucer knew well and part of which he may have translated as a young man; although there is no question of mirrors designed solely for strategic defense in that work, Dame Nature herself offers a long discussion of the diverse properties of looking glasses, a virtual citation of Roger Bacon's *Letter on the Secret Works of Art and Nature,* and references to Aristotle's and Alhazen's treatments of vision.[19] But the point here is less Chaucer's sources—which are especially numerous in "The Squire's Tale"—than his suggestion that what Wolfram, and eventually Leomarte and Raoul Lefèvre, assumed to be magical could, in fact, be brought about by nature, or "naturally." Whereas in *Parzival* the German poet had specifically ruled out the relevance of geometrical knowledge to the construction of the instrument, Chaucer appeared to follow Dame Nature's observation that the mastery of geometry was essential to the study of optics, and by implication that access to the discipline explained the proliferation of the device.

The Pharos of Alexandria

By far the most important version of the imperial mirror, and the likely source of the Virgilian and Far Eastern versions and their hybrids, would remain practically unknown to most Europeans well into the sixteenth century, when its tardy emergence offered a kind of retroactive respectability to the more familiar medieval versions. This story was associated with the Pharos, the famous lighthouse of Alexandria, and was a variant on an earlier, even more widespread, and possibly accurate assertion that the light in that tower could be seen from a distance as great as three hundred *stadia,* or thirty miles.[20]

The technological triumph represented by the Pharos would have been the correlate of the celebrated Library of Alexandria, and the various stories of the rise, decline, and fall of these monuments are to some extent related. The library, said to offer a miniaturized version of "the world," would have found in the mirror a suitable emblem of its ambitious collection. Two legends about the Pharos supplement older traditions about the library itself and anticipate one version of the destruction of that institution. An important part of the library's holdings derived from ships allowed to approach its harbor: visitors needed to surrender any books on board, and when they left Alexandria, they took away mere copies, ceding the originals to the library. Tales about the telescopic glass in the Pharos are concerned, not with these textual replicas and the diminished prestige they conferred on departing ships, but rather with the visual simulacra of approaching vessels, for only those whose mirrored image looked acceptable were granted peaceful entry. And insofar as other accounts of the device in the Pharos insist that its primary purpose was not just to spy out but rather to set fire to ships at sea, stories about the fiery fate of the library itself, or of the thousands of books kept in dockside warehouses, gain a certain inevitability.[21]

The assorted claims about the vanished mirror or mirrors in

the lighthouse from which Byzantine foes had once been surveyed arose in response to the fall of the Egyptian city, the largest in the ancient world, to Arab forces in A.D. 642–646, to its continued survival as an entrepôt for luxury goods but its political nullity, to the structural damage suffered by the lighthouse in a series of earthquakes between the tenth and fourteenth centuries, to Islam's enmity with and interest in Byzantium, and to the advances, largely theoretical, made in catoptrics throughout this period.[22]

Dating to the ninth or tenth century in Arabic accounts, the details about the mirror appeared several times in twelfth-century Spain.[23] In one such work, an anonymous Castilian translation of an Arabic universal geography, the mirror was said to reveal those approaching Alexandria from "the country of the Franks or of the Armenians or those who come from Sicily, Crete, and other Christian islands." Its form was "concave, not plane, with a diameter of seventeen cubits," and it was destroyed, like that at Rome, by foreign invaders who misled treasuring-digging local citizens.[24] Abū Hāmid Al-Garnātī, living in Alexandria around 1115, drew the Pharos and noted the telescopic powers of its mirror in two of his works, describing the device as composed of "Chinese iron" and capable of revealing and subsequently burning all naval traffic approaching from Byzantium.[25] The Spanish Jew Benjamin of Tudela also mentioned the mirror in his record of a journey to the Near and Far East from around 1159 to 1173, the *Sefer massa'ot* or *Itinerary;* made of glass, the mirror was said to be the achievement of Alexander the Great, and when "ships came to the city from Greece or the West with warlike aims, [they] could be seen when they were still twenty days' travel away, such that the inhabitants were able to mount a defense."[26] Inevitably, of course, the mirror was broken, and as a result the city's political and cultural preeminence in the ancient world came to an end.

These twelfth-century European accounts of the Pharos reflect a very strong tradition in Arabic and Persian texts.[27] As contributions to the myth of the imperial mirror they appear to have enjoyed only a very modest currency in the Latin West, though a few features seem related to the Spanish, Roman, or Far Eastern equivalents. Like that of La Coruña, for example, the optical device at Alexandria was often next to or in the hand of an animated statue.[28] When the mirror was not said to be of glass, as Virgil's was, it was frequently said to be composed of "Chinese iron," a tacit allusion to the metallic mirrors more characteristic of the Far East, and perhaps to a perception of superior craftsmanship and military might emerging from that region, but surely for some readers a reminder of the tale of Prester John.[29] And in some versions, the story of the greedy treasure hunters who were persuaded by foreigners to procure the destruction of their own tower and mirror explains the annihilation of the Alexandrian as well as the Virgilian instruments.[30]

By and large, European readers seem to have been much less well acquainted with the story about the Pharos than with those concerning Virgil and Prester John.[31] Of the four accounts originating in twelfth-century Spain, only Benjamin of Tudela's *Itinerary* was published in Europe; it was printed in the original Hebrew in Constantinople in 1543, in Ferrara in 1555, in Freiburg in 1583, and in Leiden in 1633. Its relatively late publication in Latin, in Antwerp in 1575, meant that the Alexandrian mirror became familiar to most Europeans well after its analogues from the romance tradition.[32]

The substance, function, and status of the device at Alexandria underwent many changes in the Arabic and Persian texts, and images accompanying descriptions of the device leave unclear the nature of the mirror involved. The mid-fourteenth-century account by the geographer Hamd-Allah offered an array of possibilities to explain the workings of the fabled mirror: "At the command of Alexander the Great, the philosopher Apollonius constructed a mirror seven cubits in diameter, which was placed on top of a round tower, such that it was much higher than all of the other buildings. By virtue of a talisman, when someone looked in the mirror, he could see everything that was taking place in Constantinople, despite the fact that between this city and Alexandria there is the Mediterranean Sea and three hundred leagues."[33]

The geographer's initial suggestion that the height of the mirror was crucial to its telescopic function is further complicated by his mention of "the philosopher Apollonius." Given the context, one might infer a reference to one of the greatest mathematicians of Ptolemaic Alexandria, Apollonius of Perga, whose lost work *On the Burning Mirror* would seem to allude to the tradition of immolating enemy ships, and whose discussion of types of curves in the extant *Conics* would influence subsequent work on the focal points of such instruments.[34] But many of Hamd-Allah's earliest readers would have confused this figure with Apollonius of Tyana, a Cappadocian neo-Pythagorean magus of the first century A.D. whose exploits rivaled those of Christ, and who, especially within the Arabic tradition, was the alleged "inventor of talismans."[35] The occasional association of the two philosophers—common in both the Arabic and the Greco-Roman worlds, and crucial to early modern European impressions of the mirror at Alexandria—seems to have long predated and survived Hamd-Allah. In 1625 Gabriel Naudé complained with a librarian's zeal first of the carelessness of the sixth-century monk Cassiodorus, who had vaguely identified Apollonius of Tyana as a "celebrated philosopher," and then of the sloppy scholarship of his own peers Jean-Jacques Boissard and Pierre de L'Ancre, who "say and affirm that one can still see in the Vatican Library today a book *On Conic Figures,* written by Apollonius of Tyana, the ambiguity of the name having made them mistake this man for Apollonius of Perga, otherwise known as 'the Great Geometer,' who lived at the time of Cleomedes, 150 years before the birth of Christ. It was the latter Apollonius who wrote eight books on conic figures, four of which were translated from Greek by Federico Commandino, and published in Bologna in 1566."[36]

What is important for our purposes about Hamd-Allah's account is the statement that the mirror in question had allegedly been inscribed with talismanic images and figures made under certain constellations: this detail, which reappears in several other descriptions of the Pharos, would eventually reemerge in sixteenth-century European discussions of telescopic devices.[37]

Events at Oxford

Medieval accounts of the mirrors associated with antiquity and the Far East competed with another optical instrument of legendary capabilities, the one described around 1250 by the Franciscan Roger Bacon in his *Letter on the Secret Works of Art and Nature*. The fifth chapter of Friar Bacon's work, which would first be published in Latin in Paris in 1542, translated to French in 1557, and to English in 1597, is devoted to both mirrors and some sort of magnifying lenses, the latter of which were ambiguously described as

> Glasses so cast, that things at hand may appear at distance, and things at distance, as hard at hand: yea so farre may the designe be driven, as the least letters may be read, and things reckoned at an incredible distance, yea starres shine in what place you please. A way, as is verily believed, *Julius Cæsar* took by great Glasses from the Coasts of *France,* to view the site and disposition of [both] the Castles and Sea-Towns in great *Britain.* By the framing of Glasses, bodies of the largest bulk, may in appearance be contracted to a minute volume, things little in themselves show great, while others tall and lofty appear low and creeping, things creeping and low, high and mighty, things private and hidden to be clear and manifest.[38]

The scant details given about Julius Caesar's optical device would certainly have been read as an adjustment to the more widespread legend of the Virgilian mirror, and they may have been intended as such. The instrument seems, in this account, to explain not the durability and eventual downfall of an entire empire, but merely Caesar's brief incursions into Great Britain in 55 and 54 B.C., near the end of Rome's Republican period. The silence about Virgil, who would have been twenty-five years old at this point, and about his fabled mirror suggests that either that story or the mirror itself was derivative. Moreover, the relatively rare European reader who was familiar with the legend of the Pharos might have found Bacon's anecdote convincing because the events in Gaul also preceded Caesar's expedition to Alexandria by a few years. Caesar was said to have built a tall tower, occasionally confused with the lighthouse itself, in that city, and it was sometimes suggested he had set fire not only to the port but also to some portion of the Library of Alexandria, the likely repository of crucial optical knowledge.[39] Caesar, whatever he did there, had already seen it all.

These claims aside, it is quite possible that Bacon encountered lenses—either of beryl or higher-quality ones of crystal—used to help elderly readers, but it is entirely unlikely, of course, that in the course of his experimentation and research he ever effected anything resembling what is described in the early work that is this *Letter.*[40] Similar assertions would be made in his *Opus Majus* of 1267, where it is said that "the wonders of refracted vision" allow one to "read the smallest letters and number grains of dust and sand," to magnify small bands of soldiers into large armies, and to "cause the sun, moon, and stars in appearance to descend here below."[41] The resemblance of the first part of the passage to a statement in Robert Grosseteste's *Concerning the Rainbow and the Mirror,* which alludes to the branch of optics that permits one "to read the smallest letters from an incredible distance, and to count sand, or grains, or blades of grass or any other minute object," suggests that these claims derived more from conventional expectations than from actual observation. Grosseteste's reference in turn may derive from the influence of Seneca's *Natural Questions,* where the effect was achieved not by lenses but through the use of a water-filled glass ball.[42] What is crucial here is not the inauthenticity of these statements, but rather the longevity of the expectations they engendered; chief among the desired features of the newly invented telescope was the opportunity it gave its user to spy on those reading letters some distance away.

The Baconian legacy flourished not just among subsequent students of optics but also in early modern accounts devoted to the man himself. It is not surprising that as reading glasses both improved and became more familiar, hyperbolic descriptions of Bacon's instrument increasingly came to involve the mirror rather than lenses. A late fourteenth-century manuscript related that the Franciscan's mirror for seeing "what people were doing in any part of the world" so distracted Oxford students from their work that university authorities had it broken, and in the mid-sixteenth century Robert Recorde took the legend of "a glasse that he made in Oxforde, in which men might see thinges that were doon in other places," seriously enough to defend the friar from charges of necromancy and to suggest that the feat was actually brought about by sophisticated optical devices.[43] The late sixteenth-century *Famous Historie of Fryer Bacon,* based on the extravagant claims of the *Letter,* reiterated the looking glass's ability to magnify and to contract visual phenomena, and repeated the story about Julius Caesar's view of English fortifications from Gaul.[44] The Baconian motif, because it drew upon both English pride and its not infrequent tensions with its neighbor across the Channel, would reappear as an occasional alternative to the imperial mirror throughout the early modern period, particularly in discussions of telescopic devices combining lenses and mirrors, and the notion of Bacon as the inventor of the Dutch instrument was still flourishing in the late eighteenth century.[45]

Northern Italian Optical Knowledge

Speculation about telescopic devices before the actual invention was by no means restricted to the literary arena, and public interest in the diverse functions of mirrors seems to have been strong in sixteenth-century Italy, particularly in Venice, a prominent port and glassmaking center, and in the university setting of Padua, and at the courts of Ferrara, Mantua, and Turin. As in discussions of the instruments deployed in the Pharos and by Virgil, Prester John, and Roger Bacon, explanations wavered between the rational and the occult. A number of thinkers insisted that both knowledge of optical principles and familiarity with talismans were crucial to the mirror's design.

In his *Oration on the Dignity and Utility of the Mathematical Sciences,* an inaugural lecture made at the University of Padua in 1464 but not published until 1537, the German astronomer Johannes Regiomontanus had praised Archimedes for his "philosophic" mirrors—meaning, in all likelihood, those concave in form, conventionally associated with burning distant targets—and had vowed to produce some of his own soon.[46] As far as we know, that promise was not met, but catoptrical speculation continued in Padua and seems to have focused, in at least one instance, on the relevance of Apollonius of Tyana's talisman to his mirror's telescopic potential. Thus in his *Treatise on Enchant-*

ments, composed around 1515–1520 and published in 1556, Pietro Pomponazzi, a celebrated Aristotelian and professor at the University of Padua, alluded to a discussion about the means supposedly used by the magus to see remote events. The story, as Pomponazzi recounted it, followed two others involving charlatans who had suggested that what they managed with mirrors alone actually depended upon some sort of enchantment. In the philosopher's view, natural laws of reflection, rather than supernatural means, explained Apollonius's feat.

> We can see things that are behind our backs, as happens when we place a mirror right in front of our eyes, and in this fashion, even things that are in the heavens. And I offer a story bearing witness to this: in Padua, many were in the court of the bishop—a man who was not only learned, but also very pious, Pietro Barozzi—when the conversation turned to Apollonius of Tyana, who saw things that were extremely far away. Since many there attributed this to magic arts, the most learned bishop smiled (for he was above all knowledgeable in mathematics), and said that there was nothing supernatural in this. Objects here below send their images and species into the air and all the way to the heavens, and they return again and reverberate here below, like one mirror in another. And thus such things can be seen a long way away. He added many authors who subscribed to this, the names of whom I know longer remember, and many stories about this man [Apollonius], and he said that others had been believed saints on account of such deeds, who, in view of their crimes, should rather have been judged devils.[47]

Pietro Barozzi's explanation, or at least Pomponazzi's version of it, did indeed have a predecessor in fifteenth-century discussions of optical illusions; clouds or other forms of condensation, wrote Niccolò Tignosi in his commentary on Aristotle's *De Anima,* often reflected terrestrial images from one site to another.[48] Such accounts evidently possessed a certain plausibility and longevity: in his study of talismanic art of 1629, Jacques Gaffarel proposed atmospheric reflection as a means of "conveying news in less than an hour to someone more than one hundred miles away."[49]

However unsatisfying Pomponazzi's solution might be—for it neither corresponds to general experience nor makes clear Apollonius's special exploitation of the phenomenon—it does suggest that the legend of the mirror at Alexandria was under serious discussion in the Paduan milieu, possibly at the university, where Bishop Barozzi served as chancellor and Pomponazzi himself was a professor from the late 1480s until 1507. It also implies that for a number of their interlocutors, no real distinction was made between the talisman upon which this particular mirror allegedly depended and the physical laws governing the behavior of all such glasses. Despite the presumed objections of thinkers like Barozzi and Pomponazzi, the conflation of potentially supernatural and purely technical impressions of the mirror would survive.

An important modification of Pomponazzi's view arose in the work of one of his students, Girolamo Fracastoro of Verona. In his *Homocentrica,* first published in 1538, Fracastoro made refraction, not reflection, a crucial factor in the everyday observation of the heavens. It was not that images of terrestrial objects reverberated from earth to the sky and back again, but rather that differing densities in the airy medium between the observer and the heavens distorted celestial objects. The stars alone, being composed of denser material, were compared to mirrors.[50] Interestingly, once Fracastoro had discarded both the occult and the natural versions of the telescopic mirror, he introduced several important references to the use of glass lenses for distance vision.

> If someone looks through two eyeglass lenses placed one on top of the other, he will see that everything looks much larger and closer . . . If a lens is placed in the medium between the object and the eye, the object will look much bigger. If, however, the lens is right next to the eye or the object, it will look much smaller.[51] Some eyeglass lenses are made of such thickness that if someone were to look through them at the Moon or at another star, he would imagine them so close that they would seem no higher than the towers.[52]

Fracastoro's assertions were made in the service of an argument associating magnification with a dense medium, and it appears that he was discussing the use of convex lenses. His statement about the combination of lenses was not entirely accurate.[53] It did, however, earn him an early but rather tenuous place as an inventor of the telescope, not only in Galileo's lifetime, but even in the late eighteenth century.[54]

Fracastoro's biography, written by his friend Giovanni Battista Ramusio and published in 1555 as the preface to the *Opera Omnia,* did not mention any particular optical devices but perhaps kept alive the legend of some sort of telescopic vision in its depiction of his villa. Located on an elevated spot some fifteen miles from Verona, Fracastoro's home was said to be a site from which "the city, and at a great distance, the innumerable surrounding farms in the neighboring plain could be seen, and within these places, wandering herds of all sorts, and smoking roofs around evening," and likewise "the high waves of Lake Garda, and the prosperous peninsula of Catullus, and also fleets of sailing vessels, and fishing ships approaching from afar."[55] Fracastoro himself was curiously described as deformed by some sort of viewing apparatus—probably a Jacob's staff rather than an instrument with lenses—"his nose being snub and compact because of his constant contemplation of the stars."[56]

Ramusio, like Fracastoro, appears to have his place in an intellectual context where the properties of mirrors and lenses were being compared, and where some measure of skepticism colored traditional claims about the former. In 1550 Ramusio had published a work that would have modified what little Europeans knew of the Pharos, the best-selling *Description of Africa* by the Grenadan convert Al-Hasan al Wazzan or Leo Africanus. Emerging as the first in a collection of travel narratives dedicated to Fracastoro, the story was introduced as a legend, and the device in question was a large steel mirror mounted on top of a high column. The city of Alexandria was defended from approaching vessels not necessarily—or at least not solely—because of any telescopic instrument, but rather because "all ships passing near the column when the mirror was uncovered were miraculously and instantly burnt up."[57] Leo ended this account—which may be a condensed version of the kind offered by Abū Hāmid Al-Garnātī, where immolation followed briskly on telescopic revelation—with a nod to regime change and above all to the dubious nature of the tale. "They say that the Muslims ruined the mirror, such that it lost its power, and they took away the column: a truly ridiculous thing, and something to tell children." In a remarkable variant, ascribed to a manuscript untouched by Ramusio's heavy editorial hand, Leo seems to have presented the mirror entirely without skepticism: it was a wondrous invention of "a philosopher named Ptolemy," damaged by the Muslim invasion, and wrecked, finally, by "a Jew who rubbed it with garlic."[58]

Whatever Leo's beliefs about the mirror, the numerous translations of his *Description* conform to the skeptical account offered in Ramusio's edition. But mid-sixteenth-century sources suggest, in the meantime, that northern Italians were well acquainted with occult explanations of telescopic mirrors, particularly those involving talismans. Consider in this connection, for example, a tale told by the Benedictine abbot Johannes Trithemius and repeated by the Carthusian monk Laurentius Surius:

There appeared in those days [in 1501] in Lyon, France, an Italian man named Giovanni, who preferred to be called "Mercury" on account of all sorts of ancient learning that he displayed. He went around with his wife and children, all clad in linen, and wearing, in imitation of what we are told by Damis about the philosopher Apollonius of Tyana, an iron chain about the neck. He promised great things, and he gloried in having more learning than did all the ancient Hebrews, Greeks, and Romans . . . He was admired for some time by the king of France [Louis XII], upon whom, it is recalled, [Giovanni / Mercury] conferred two overwhelming gifts. One was a sword made of 180 small daggers, and the other a shield decorated with a marvelous mirror. He said in a certain book that these two miraculous things were made in a miraculous manner under a certain constellation.[59]

Although the ultimate source of such mirrors would appear to be the legendary Apollonius of Tyana, the tale of the odd man who was Giovanni Mercurio da Correggio suggests that the place to which occult knowledge of this sort had been transferred is Italy, and that the ultimate political beneficiary of this learning was France.[60] This is precisely the pattern of transmission that will reemerge, as Chapter 4 will show, in a pamphlet published in November 1608 about a telescopic mirror owned by Henri IV of France.[61]

Explicit indices of interest in the telescopic properties of the fabled Alexandrian mirror emerge in this period in the northern Italian city of Ferrara. In 1582 the *Compendious Introduction to the First Part of Catoptrics* was published by the otherwise unknown, and perhaps pseudonymous, Rafael Mirami, who identified himself as a Jewish physician and mathematician of Ferrara. It offered the usual allusions to trick mirrors and referred in passing to "that one that they write used to be in the top part of the pinnacle of a tower, in which men saw very distinctly the ships that were coming into port, together with all the people, and merchandise, that was uncovered."[62] Mirami's allusion was a hybrid description of both the Pharos—where the point was the approach of very remote ships—and a similar, much less celebrated, knockoff in the Tunisian port of Goletta, where the distance was considerably less and the emphasis was on docking ships, attractive goods, and the colorful minutiae of people's clothing.[63] The fact that Benjamin of Tudela's *Itinerary,* a crucial account of the Pharos, had been published in Hebrew in Ferrara in 1555 perhaps explains Mirami's rather casual tone and his willingness to conflate two different stories. It is also noteworthy that he portrayed the uses of the mirror in commercial rather than military terms, and chose not to insist on the spectacular distances at which the instrument might function: the device appears less the fabulous emblem of fallen empires than a useful tool for any mercantile state.

This is not to say, however, that Mirami was wholly uninterested in the literary preexistence of the sorts of mirrors whose form and function he described in the *Compendious Introduction:* he frequently cited, in the course of his theoretical arguments, illustrative verses of Horace, Dante, and Petrarch, and proudly compared, in an analogy made more plausible by the weird story of Giovanni Mercurio, Archimedes' burning mirror to a magician's polished shield in Lodovico Ariosto's romance epic *Orlando Furioso.* It is not surprising, then, that in his *Universal Marketplace of All the Professions in the World* of 1583, Tomaso Garzoni relied on Mirami's brief and businesslike allusion to the Pharos but sought to restore something of the hyperbolic quality it and its Virgilian and Far Eastern equivalents had once enjoyed. "And the one that they write used to be in the top part of the pinnacle of a tower, in which men saw very distinctly the ships that were coming into port, together with all the people and merchandise, that were there," Garzoni noted, "was marvelous [*fu meraviglioso*]."[64] In sum, Garzoni and Mirami alike are characterized by their readiness to countenance fabulous literary descriptions of mirrors, the best examples of which are those of Apollonius, Alexander the Great, "King Ptolemy," Virgil, Merlin, Hercules, and Prester John, as feasible prototypes of actual or potential instruments. Mirami's vow "to describe the means of manufacturing all the miraculous mirrors that have already been made by others, along with the uses to which they might be put," is offered very much in this spirit.

To the Lighthouse

References to the telescopic device in the Pharos emerge in travel literature of the early modern period, and a comparison of the observations of a number of northern visitors to the Near East is instructive. Hans Schiltberger, Hans Christoph Teufel, Georg Christoph Fernberger, Reinhold Lubenau, and Nikolaus Radziwill all mention the Pharos in their travel accounts, and their various allusions suggest an increasing familiarity with the older claims about the mirror and its eventual destruction. Their remarks range in tenor from enthusiastic endorsement of the tales they encountered to a cautious skepticism.

The Bavarian Hans Schiltberger, whose odyssey as a captured crusader in Europe, Asia, and Africa lasted from 1396 to 1427, related that in a tower near the port of Alexandria "not long ago" there was mirror allowing one to see toward Cyprus "those who were on the sea; and whatever they were doing." His story of the mirror's eventual destruction reflects his position as onetime prisoner of the infidels, and perhaps the sheer length of his captivity as well. As Schiltberger told it, a priest from Cyprus, after gaining permission from the pope to pretend to abjure, went to Alexandria, converted, "learnt their writing," and became a trusted Islamic cleric. Eventually given a choice of any mosque in the city, he selected the one equipped with the mirror, and after nine years of pious fakery he summoned at last the warships of the king of Cyprus. As his allies were approaching, he struck the mirror three times, and though the object broke, the noise of his blows alarmed the citizens, and the trapped priest perished when he jumped from the tower into the sea.

The king of Cyprus did, however, take the city, and with this allusion to Peter of Lusignan's capture of Alexandria in October 1365, Schiltberger narrowed the gap between what he portrayed as a functioning and non-necromantic mirror—albeit one of questionable substance and design—and the moment of his visit to Alexandria to less than four decades.[65] His narrative, first printed in the vernacular in 1460 and repeatedly reissued until the early seventeenth century, was familiar and perhaps credible to many readers of German. When Johannes Kepler sent Galileo's newly published *Starry Messenger* in the spring of 1610 to his friend Johannes Papius, professor of medicine at the University of Tübingen, the latter replied with cryptic brevity, "Thank you for the two copies you sent. Have you ever seen the book of Schiltberger about this kind of optical instrument, which they say survived in memory more than two hundred years ago?"[66]

Hans Christoph Teufel's impressions of the Pharos, by contrast with those of Schiltberger, seem more informed by his education than by his actual experience in Alexandria. An Austrian nobleman who had studied at the Universities of Padua, Bologna, and Siena in 1585–1586, the young Teufel left Venice for three years' travel in the East in 1587.[67] He was accompanied on the earlier part of his voyage by Georg Christoph Fernberger, who eventually went on without him to the Far East; both men produced *post eventum* accounts of their journeys. Teufel's version, originally written in German but published in Italian in 1598, shows that even prior to his arrival in Alexandria in September 1588, he was concerned with the question of the mirror:

> [From Rhodes] we entered the Port of Alexandria on the evening of the 19th [of September]; from here to Constantinople it is 1200 miles. At our entry in this port, we saw on the right hand a stronghold which is where, long ago, the Pharos was, that tower with enchanted mirrors that I have already mentioned, mirrors in which one could see, fifty days in advance, the arrival of all armies that approached with the intention of harming the kingdom of Egypt. This was the reason for which the Egyptians, seeing danger arranged so many days in advance, were able to avoid, thanks to their previsions and preparations, all armed attacks. Finally, however, the guardian of this mirror was intoxicated by a Greek and fell asleep; the Greek broke the mirror, and because of this deed the Kingdom of Egypt subsequently fell into the hands of other people, the mirror having lost its power.[68]

Though the story about the drunken guardian of the mirror had appeared in Benjamin of Tudela's *Itinerary*, other aspects of Teufel's version of the Pharos are more puzzling in origin. First of all, he seemed initially to suggest, for reasons not entirely clear, that more than one mirror contributed to the telescopic effect,[69] and impressions of this sort recall the discussion of reverberating images mentioned several decades earlier by Pomponazzi in his

Treatise on Enchantments and pursued well beyond Padua. The feasibility of combining plane mirrors to relay images of distant events was in fact a common topic of discussion in mid-sixteenth-century Latin and Italian works, as Chapter 2 will show, and in his *Theater of Mathematical and Mechanical Instruments* of 1578, the Huguenot exile Jacques Besson had suggested that a plane and a concave mirror could be used together, but he restricted this to objects at close range, for the arrangement was an aid to readers.[70] The notion that the Pharos involved more than one mirror was, however, taken up anew in 1632 by Galileo's disciple Bonaventura Cavalieri, who argued that "if we combine the concave [mirror] with the convex mirror or with a concave lens we should get a telescopic effect, and such, perhaps, was Ptolemy's mirror."[71]

But if anything in Teufel's detail about multiple reflectors indicates a sense of how the legendary feat might actually have been managed, his position is complicated by the reference to the "enchanted" quality of at least one of the mirrors. What would seem a discrepancy for us—inclined as we are to accept either a plausible physical explanation or the premise that the instrument is in its entirety magical and thus fictional—evidently did not obtain for Teufel, nor for the writers in Arabic and Persian who had discussed both the structure and the talisman upon which the mirror depended. There is even something of an analogue in the development of a novel imported to early modern European culture from Persia.

In the Italian translation of *The Travels of Three Young Men, Sons of the King of Serendippus*, published in Venice in 1557, there is some discussion of the so-called "mirror of justice," the shape [*forma*] of which had been discovered by ancient philosophers of a certain empire. Rather than showing external menaces to the state, this mirror revealed evildoers at home, and thus ensured civic tranquility: "Everyone contented himself with his station in life, devoted himself to farming, and all things flourished."[72] When the narrative of *The Travels* was adapted in 1610 by François Béroalde de Verville in his *Journey of the Fortunate Princes,* the "mirror of justice" enjoyed the same pedigree from "ancient philosophers," and performed the same function, but was also explicitly said to depend upon talismanic magic. Familiar with the properties of concave looking-glasses from his early edition of Jacques Besson's *Theater of Mathematical and Mechanical Instruments,* in this context Béroalde evidently saw the detail about the ancient philosophers' discovery of the *forma* of the "mirror of justice" as an allusion, not to the crucial issue of shape alone, but also to whatever was inscribed on the surface of the instrument.[73]

It is feasible that Teufel's expectations regarding the mirror at Alexandria were a result of his stay in northern Italy; he had left home as a seventeen-year-old, had departed for the East before his twentieth birthday, and had studied at the University of Padua in 1585 and at the Universities of Bologna and Siena in 1586, in any of which settings he might have been exposed to the latest speculation about mirrors.[74] Even though we cannot establish that Italy was the origin of Teufel's interest in the mirror, it is worth noting that he directed his travel account to Italians rather than to readers of his native German, and it is especially interesting to compare his apparent faith in the Pharos to the more cautious observations made by his companion, Georg Christoph Fernberger, who arrived with him in Alexandria aboard the same ship. "The land, which has the castle on the right side," Fernberger noted, "was once an island called the 'Pharos,' in which a most lofty tower had been constructed with marvelous ingenuity. It was said that a mirror had been placed on its summit a long time ago by Alexander, one in which all ships could be seen more than five hundred miles away, and it was reckoned among the seven wonders of the world."[75]

Fernberger's account, like Teufel's, depends in some part on the great humanist Martin Crusius's discussion of the Pharos in a work of 1580, itself a repetition of Benjamin of Tudela's recently translated *Itinerary.* Crusius, like Kepler's friend Johannes Papius, was at the University of Tübingen, but the version he offered differs somewhat in spirit from that of the captured crusader Schiltberger. In one of his annotations Crusius had written of ancient Alexandria,

> In this place a most lofty tower was built, which the local people called *Magraah,* and in Arabic *Magar Alecsandria* is *Alexandriae Pharon.* It was said that a glass mirror had been placed on its summit a long time ago by Alexander, one in which all war ships that were sailing to harm Egypt, either from Greece, or from anywhere in the West, could be seen, and guarded against, even fifty days' journey away, that is, at a distance of more than one thousand Persian leagues, and a defense could be prepared. At length, many

days after the death of Alexander, when the Greeks were under the Egyptian yoke, a crafty man named Sodorus who had sailed from Greece broke the mirror, its guardians having been lulled to sleep by drink.[76]

Fernberger, unlike his sources Crusius and Benjamin, insisted upon neither the glassy substance of the mirror nor the means of its eventual destruction. And where Teufel sought to explain the legend of the Pharos by references to multiple mirrors and to magical qualities, Fernberger supplemented the conventional allusions to the lofty tower with *miro artificio constructa,* "constructed with marvelous ingenuity," as if the building itself, rather than any of its optical components, were more worthy of note.

Fernberger's ambiguous description of the Pharos is echoed by yet another visitor to Alexandria in the fall of 1588; the Prussian Reinhold Lubenau also portrayed it simply as "of ingenious construction," plainly unwilling to commit himself to detailed speculation about the legendary mirror.[77] And the Polish nobleman Nikolaus Radziwill, who had studied with Crusius as a young man and was perfectly ready to note other marvels in his trip to the Holy Land in the early 1580s, could do no more than allude vaguely to the common historical judgment that the Pharos had been constructed "through human ingenuity, and with great expense and artifice."[78] This reluctance to offer specific information about exactly what made the lighthouse one of the wonders of the ancient world is likewise characteristic of a Turkish miniature of this same period, for there the thing on top of the Pharos, closely contemplated by two officials, is a brilliant golden sphere, an ambiguous object that hovers between a polished spherical

شبك منارهٔ اسكندر

Turkish miniature of the Pharos of Alexandria, ca. 1582, from the workshop of Ustad ʿOsman. Bibliothèque Nationale de France, Oriental manuscripts, suppl. Turc 242, fol. 76v.

mirror for detecting ships from Constantinople, and a stylized fire of use to any passing vessel and of no particular defensive value.

Travelers who confronted the wreckage of the Pharos of Alexandria seemed increasingly to see it as the emblem of an irrecoverable past. "The entire city is full of piles of rocks and ruins, a true mirror of the instability of the world," wrote the Franciscan Antonius Gonzales in the course of his travels in the winter of 1665, making Alexandria not the powerful surveyor of its neighbors but a sorry spectacle for all onlookers.[79] And in a facetious collection of French stories published in 1597, the telescopic mirror, "adorned with diabolic magic from Toledo," was wantonly transferred to the neck of the Colossus of Rhodes, such that those on the island "used to be able to see ships going to Syria, or to Egypt."[80] But many armchair travelers appeared ready to believe that with the requisite combination of ingenuity, expense, and artifice, something very much like the Pharos was within their grasp. The project of manufacturing telescopic mirrors preoccupied both Italian and English natural philosophers of the mid- to late sixteenth century. That project is the subject of Chapter 2.

Chapter Two

————◈————

Idle Inventions

$\mathcal{T}$HE interest in telescopic devices was doubtless enhanced by actual developments in the design of metal and glass mirrors, and by hyperbolic descriptions of things that once had been and might again be accomplished with such instruments. *Lesensteine* or "reading stones," transparent stones that provided slight magnification to the page beneath them, appeared around the mid-thirteenth century, and convex lenses were adapted as reading glasses shortly thereafter.[1] Both devices, especially the latter, provided a growing segment of the European population with some familiarity with the magnification of nearby objects, but when the same effect was achieved through the use of concave metallic mirrors, that much rarer experience would be described as just one of many catoptric wonders, or in terms that evoked the more spectacular dream of telescopic vision.

Thus the second part of the influential *Romance of the Rose,* which emerged around 1280 or more or less when eyeglasses did

and at least several decades after transparent "reading stones" became common, presents the mirror, rather than any glassy magnifier, as the marvelous object. The concave mirror in particular
permitted an odd kind of distance vision: the poet had more to
say about the unfathomable remoteness of the viewer than about
the magnified appearance of the minuscule writing and grains of
sand.[2] And even those who attempted to describe the minute objects under view often resorted by default to language emphasizing instead the distance of the observer: when Giovanni Rucellai
wrote of a concave mirror's magnification of the body of a new-
born bee in 1524, for instance, he began by comparing that creature to monuments or monsters traditionally viewed from afar—
the Colossus of Rhodes, a projected sculpture on Mount Athos, a
dragon from the Libyan desert.[3] The remoteness of the vantage
point, in other words, serves as a prelude to and occasionally as a
substitute for an insistence on realistic details of magnified objects.

The growth of the trade in corrective eyewear must also have
increased the distinction between what might be accomplished
with lenses and mirrors. Those who sold eyeglasses were taxed
with a kind of clowning hucksterism—in a very popular work of
about 1515, for example, the wandering fraud Till Eulenspiegel
pretends to be an unemployed spectacle maker from Brabant,
and he dresses in odd clothes for the occasion—but theirs was a
quotidian trickery associated with the relentless hawking of defective products to gullible consumers.[4] The claims made about
mirrors, less easily tested, were often much greater in scope.

Glasses for Seeing Far

One of the earliest references to a functioning telescopic mirror occurs as a laconic observation in the correspondence of Charles de Marillac, a French ambassador in London in the spring of 1541. Marillac reported on the instrument at Dover Castle during a period of tension between France and England, but unlike the man who appears to have paid for it, he seemed indifferent to the possibility of the mirror's deployment. "Around here there's an Italian who must be about seventy years old," Marillac related, "and he has given the king [Henry VIII] to understand that he can make a mirror and place it on the summit of Dover Castle, and that looking into this mirror one would be able to see all the ships leaving from Dieppe. And though this seemed an incredible thing to many, he so persuaded the aforementioned lord that he was given a supply of money to do this, and in fact he left here yesterday to go to Dover to carry out what he promised."[5]

What is striking about the story is that while it promises a technological novelty, it also conveys something familiar, worn, and even banal. Early modern onlookers would have seen the inventor's proposal as a mirror image of the story told about Julius Caesar's survey of Britain from Gaul, for the vantage point and target have merely been reversed. Of this particular old Italian and his mirror one hears no more, and for many the whole enterprise must have seemed just one more expenditure in the costly but fruitless efforts to turn the castle and the disastrous silted-up harbor at Dover into viable defense systems. It is plausible, moreover, that the would-be inventor had simply appropriated and updated the claims about another ancient mirror made by a late compatriot, the Sienese engineer Vannoccio Biringuccio. The latter's *Pirotechnia,* written around 1535 and posthumously published in Venice in 1540, includes a tale about an antique mirror once said to have been in Tunis, a device that revealed the ships and the people in the nearby port of Goletta, "and in what colors and clothes they were dressed."[6]

The brazen claims of Marillac's old Italian would soon find their echo elsewhere in mid-sixteenth-century Italy. Magnifying mirrors were a particular topic of interest, and in the 1567 edition of his encyclopedic survey of various professions and arts, Leonardo Fioravanti concluded his description of mirror making with a reference to Ettore Ausonio, a physician, alchemist, and mathematician in Venice, and the acknowledged master of the trade in looking glasses:

> I saw in the celebrated city of Venice mirrors that were miraculous in their functions. They were made by that great philosopher and mathematician Ettore Ausonio of Venice, an inventor of the most splendid mathematical devices that have ever been seen or heard in the world. He had made several concave mirrors of incalculable size, in which one might see marvelous and enlarged images, which I will not describe here, since by now all the princes of the world are familiar with them. And in addition to mirrors this extraordinary man developed so many wonderful things relating to mathematics, that it is a miracle, as I will explain more thoroughly in the chapter on mathematics. And this mirror of his, as I have said, is so amazing, that those who see it are stupefied. If I were to write of all the strange and bizarre things I

saw in this one and in others made in different ways, and recount all their curious effects, it would be endless, and even with all that I might say, it would finally be nothing [compared to the actual phenomena].[7]

The breathless quality of Fioravanti's description notwithstanding—he was called "nothing but an Italian charlatan" by René Descartes[8]—Ausonio did write an important treatise on spherical concave mirrors around 1560, and in this same period he supplied a variety of mirrors, crystal lenses, and mathematical instruments to at least one princely patron, the ambitious Duke Emanuele Filiberto of Savoy. And that Ausonio was regarded as an authority in the ongoing discussions of telescopic vision can be inferred from several documents from the 1550s.

At some point in this decade, while Ausonio was associated with the prominent Venetian printer Michele Tramezzino, an unidentified correspondent sent him a letter about "how to see within a room, by means of a mirror and perspective, everything that is done far away in a piazza or elsewhere." The procedure, obscurely described and accompanied by a rough drawing, involved two towers, one surmounted by a large plane mirror and the other outfitted with an observation room. The mirror, or in the case of a very sizable piazza, several plane mirrors or a single convex one, would be trained on the "festival or carousing" and the image of such activities was then to be reflected to the observer in the room of the second tower. Though the observational chamber is not described in any detail, it might have been outfitted with a second mirror: this much is perhaps suggested by the writer's reference to "the way to arrange the *first* mirror."[9]

The setup does not seem a feasible one, and though he pre-served the letter, it is unlikely that Ausonio was very impressed by its contents. What is notable, however, is its generic resemblance to other discussions about how to convey mirrored images over relatively small distances to observers placed in convenient towers or well-situated observation rooms. The letter recalls, in fact, two unlikely procedures described in 1550 by Girolamo Cardano in a weird pairing of prurient activity and military reconnaissance. In Cardano's *On Subtlety,* readers were advised that they could see ei-ther "whatever is being done in yonder bedroom, provided there is light," or across a distance of five miles, and over walls and into enemy territory, if they had an elevated plane mirror trained on the site of interest and a second, handheld mirror into which that image might be reflected.[10] And although the 1560 edition of Cardano's work would bear the caveat that any of the images from enemy territory would be so reduced in size as to be visible "solely to those with the sharpest sight," in 1558 the young Nea-politan Giambattista della Porta, like Ausonio's correspondent, had been indifferent to such obstacles and had presented a peri-scope-like device as if it were capable of showing "what is done far away, and in other places."[11]

Consider, by way of contrast, a second letter addressed to Ausonio in this period. The author was Francesco Angelo Coccio, a well-known translator of Greek and Latin texts, and he wrote from Treviso, about fifteen miles from Venice, in 1557.[12] Coccio began his letter with the confession that he had no news, and there does seem to be absolutely nothing to report from his tem-porary home: "If you want to see that I have nothing to write, here, I am putting down my pen, and I am taking up my glasses for seeing far [*occhiali da veder lontano*], and without getting up from the table, I see Venice in the way that one sees certain distant cities represented in the landscapes of Flemish painters, and if the body had the spirit's ability to travel with the mind where it pleases, you would see me in person next to you at this very moment in which I am writing, but since this can't be done, I have left this business to my mind."

In some sense Coccio's odd figure of speech is a typical display of nostalgia and requires no optical props: a millennium and a half earlier, the exiled poet Ovid, writing in the final years of his life from the Black Sea, had likewise offered elaborate vignettes of all of Rome that remained in his mind's eye.[13] More notable is the fact that Coccio began by envisioning the Venetian landscape but ended by imagining his own likeness at his friend's side in that city, as if the telescopic instrument were simultaneously available to both correspondents, or more prosaically, as if it were an idea familiar to each. Venice itself, moreover, undergoes a sort of generic transformation in Coccio's letter and becomes simply one of those remote cityscapes depicted by Flemish artists. Put differently, it is less a physical place at a certain distance from Treviso than a style of pictorial representation then well known in the Veneto.[14]

Despite the unreal quality of the "glasses for seeing far," however, there is reason to believe that Coccio had a general notion of telescopic vision and that he, though not necessarily Ausonio himself, imagined that such an effect was produced with glass lenses rather than with mirrors. The letter ends with a request that Ausonio remember Coccio to two particular friends, the painter Giuseppe Salviati and the sculptor Danese Cattaneo, either of whom might have shared their interest in the optical issue. Salviati, known as "painter and mathematician," had served just a year earlier as illustrator for Daniele Barbaro's translation of and commentary on Vitruvius's works, a context in which the use of "an old person's eyeglass," or a convex lens, was eventually adapted to the camera obscura to produce miniaturized views of exterior landscapes.[15] And Salviati's close associate Cattaneo, portrayed by Coccio as excessively busy in those days, was in fact engaged in producing a statue of Girolamo Fracastoro as a modern astronomer holding a celestial sphere.[16] Fracastoro's references to convex lenses in his astronomical work figured as a notable alternative to the usual insistence upon mirrors as the crucial feature of telescopic sight, and his recent biography, consulted by Cattaneo in the course of his sculpting, presented the panoramic views from Fracastoro's villa to Verona, likewise fifteen miles away, in terms that were surely hyperbolic but entirely consonant with such vision. It is possible that Coccio imagined that the effect provided by his "glasses for seeing far" could in theory be produced by lenses.

The foregoing suggests, then, two competing notions of telescopic vision, the first involving the projection of images across a certain rather limited space from one mirror to another, and the second the magnification of images through the use of convex lenses, possibly within a dark room. These procedures would not have produced satisfactory results over any significant distance, but they might have been seen as alternatives of a sort. It is worth noting in this connection that Girolamo Cardano, directly after his bold suggestions that mirrors could be used to convey images from "yonder bedroom" and enemy territories, offered a one-sentence description of the camera obscura with a convex lens, limiting the range of its functions and, most importantly, presenting it in the midst of pages upon pages otherwise wholly concerned with mirrors: "If you desire to see what is being done in the street, while the sun is shining brightly, you will fix a round globe of glass in the window, and then when you have covered up the window, on the opposite wall you will see images brought in through the aperture, but they will be dim in hue; therefore, set up a very white piece of paper in the spot where you see the images, and you will manage the intended effect admirably."[17]

For his part, Ausonio recommended the use of the concave spherical mirror within the camera obscura, albeit one without a convex lens in the aperture. In manuscript notes dating to about 1560 and devoted to effects produced with concave mirrors, he pointed out that "if a man encloses himself in a darkened chamber, and arranges it so that sunlight enters only through a tiny opening, and above this opening he places a white paper, by adjusting the mirror a certain distance from the aperture he will represent on the paper a beautiful and distinct miniature painting that will place everything that can be seen outside in perspective. And if there are men or animals that are moving outside, they will move within the picture, to our immense delight."[18]

The concave spherical mirror would have projected a sharp and bright inverted image, but one whose left–right reversal had been corrected. There is some evidence to suggest that Ausonio thought of this procedure as an improvement upon the lens-based one described by Cardano, for his discussion of the camera obscura, like the one in *On Subtlety,* is immediately followed by an allusion to the use of concave mirrors to show spectators their inverted images or their faces adorned with only one eye or grotesquely enlarged.[19] And in a passage preceding his description of the camera obscura, where Cardano had offered hints about how to deploy mirrors to scrutinize activity in nearby bedrooms provided they were well lit, Ausonio boasted of his means of illuminating such areas. "This summer some people were sleeping with windows open and without light so as not to be seen," he wrote in a passage presumably never intended for publication, "and I took up my [concave] mirror, and with a candle I sent light into the neighbors' bedchamber, and I saw the room and those who were in it as if it were daytime."[20]

The discussions associated in this period with Ausonio—which Coccio described, somewhat disingenuously, as "full of various teachings and wise learning, sometimes mixed with honest pleasures"—seem in sum to have involved comparing and contrasting the optical features and effects provided by lenses and mirrors, but not combining them in a single device. Such combinations would, however, undergo discussion in the following decade within the English context, in developments inevitably presented as outgrowths of instruments long ago deployed by Roger Bacon or in the Pharos.

Ausonio's legacy, finally, lay in his unpublished *Theoretical Discourse on Concave Spherical Mirrors,* a manuscript variously known

to Paolo Sarpi, Galileo, and Giovanni Antonio Magini, and probably to Giambattista della Porta as well.[21] Written around 1560,
Ausonio's work offers a factual and visual presentation of the
principal phenomena associated with concave mirrors, and thus
lists in a schema of possibilities the mirror's use in bright sunlight
in activities such as heating the air and thus burning, cooking, or
melting certain objects, and in projecting writing onto remote
screens; its deployment in conjunction with other forms of illumination to produce visual images of varying size, orientation,
and location; its ability, in any sort of lighting, to propagate
sound and generate echoes, and finally its utility in projecting
within the camera obscura either nearby sunlit phenomena, or re-
moter nocturnal scenes in enemy camps and letters illuminated
by candles or torches.

Two aspects of the *Theoretical Discourse* are especially significant. Ausonio showed the five possible locations of images
formed by the concave spherical mirror, not all of which were
within the conceptual range of medieval and early modern optics.
Moreover, because his work also addressed the mirror's utility in
combustion, his presentation showed the relative proximity of
the place where targets would burst into flame most quickly, and
the point where images, upright and at their greatest enlarge-
ment, begin to collapse and to undergo inversion and contrac-
tion. Put differently, Ausonio's schema illustrated a feature not yet
known—the focal point—by treating the projects of burning
things and of merely seeing their enlarged image under the same
rubric. To oversimplify, the theoretical framework he offered was

a sober version of the extravagant legends of the Pharos, where a mirror set approaching ships on fire, or kept them under watch, or both.

Lively Images

The Baconian legacy, although not unknown on the Continent, was best developed in England. The late sixteenth-century *Famous Historie of Fryer Bacon,* based on the exaggerated claims of the *Letter on the Secret Works of Art and Nature,* reiterated the glass's ability to magnify and to contract visual phenomena and repeated the story about Julius Caesar's view of English fortifications from Gaul.[22] And *The Famous Historie of Fryer Bacon* also inspired Robert Greene's *Honorable Historie of Frier Bacon and Frier Bongay,* a play of about 1590 in which a "glass perspective" was the means by which protagonists viewed seemingly remote events, and a clever substitution for the letters in which such actions were normally summarized.[23] Although part of the humor of the optical prop is that it was deployed to observe matters only a few feet away, and thus rendered nearby objects distant, within the logic of this play the device itself was no sham and had eventually to be destroyed precisely because of its excessive capabilities.

Four English individuals were especially associated with the Baconian instrument: the well-known magus John Dee of Mortlake, Leonard Digges of Kent and his son Thomas, and the mathematical practitioner William Bourne of Gravesend. To judge from a remark made in a meeting of the Royal Society late in the seventeenth century—one in a litany of allusions to early, unrecognized, and mostly English inventors of the telescope—Dee had written a commentary on Bacon's *Letter* mentioning the natural, rather than occult, means by which the Franciscan had viewed distant objects. That text seems not to have survived, and in this decade Dee might have been more interested in the combustion provided by certain concave mirrors than in their telescopic features. He returned to the issue in 1570, however, in an influential preface he wrote to an English translation of Euclid's *Elements:* in a general discussion of geometrical devices, he noted that a military commander might "wonderfully helpe him selfe, by perspective Glasses," to make an accurate estimate of enemy forces and equipment, but he gave no details about the instrument and appears to have left its refinement to "posterity."[24]

Dee did in any case own a large concave mirror, and both this instrument, typically described as a "perspective," and hyperbolic tales about it survived him. Upon Dee's death in 1609, the glass itself went to Thomas Allen, an influential teacher of mathematics and antiquarian whose manuscript collection included Bacon's work. Dee's "perspective," like Bacon's glass, was also a familiar stage prop: it appeared as a thing that worked rather well in a drama written between 1597 and 1602, and as the dubious possession of a foolish philosopher who had "solde all for a glasse prospective" in a satire of 1608.[25]

In 1571 the young Kentish gentleman Thomas Digges, who regarded Dee as his "revered second mathematical father," published a slightly more explicit description of "perspective glasses" in *A Geometrical Practice, Named Pantometria,* a work written for the most part decades earlier by his late father, Leonard Digges.[26] The miraculous manner in which the combination of a glass lens and a mirror revealed the features of remote landscapes, especially of relatively flat countryside, is presented as the corollary of a standard Euclidean procedure in which a plane mirror was used to gauge the distance of certain objects.

The basic problem was the measurement of the height of an inaccessible object when the sun was not bright enough to cast shadows. Rather than establishing the ratios between a shadow of an object whose height was known, typically a gnomon or stake placed in the same sight lines, and that of the inaccessible object, the surveyor used the reflected image of the remote object, the rectilinear rays on which Euclidean optics is based, and the equal angles of incidence and reflection to establish two similar triangles whose proportional relationship could then be calculated. The procedure with a plane mirror, for which a water-filled basin could sometimes be substituted, is described in Euclid's *Optics* but seems to have come from the *Catoptrics.*[27] That technique, a variant on analogous problems in early modern military manuals and perspective treatises, allowed anyone who knew the height of the cliff or tower where he was located to calculate the distance of an approaching ship, for instance, by using the equal angles of its reflection in a mirror to construct two similar right triangles.

Digges followed his exercise in altimetry, which would work

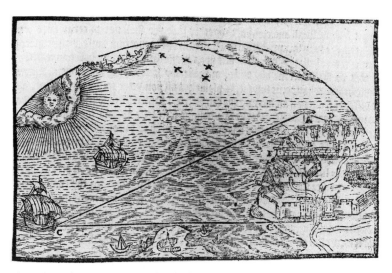

Altimetry with a mirror; Leonard and Thomas Digges, *Pantometria* (London: Abell Jeffes, 1591), 27. Grenville Kane Collection, Rare Books Division, Department of Rare Books and Special Collections, Princeton University Library.

best if the ground were flat, the cliff sheer in its drop, and the ship anchored, with a more dramatic disclosure about telescopic vision.

> This muche I thought good to open concerning the effects of a playne Glasse, very pleasant to practise, yea most exactly serving for the description of a playne champion countrey. But marveylouse are the conclusions that may be performed by glasses concave and convex of circulare and parabolicall formes, using for multiplication of beames sometime the ayde of glasses transparent, whiche by fraction should unite or dissipate the images presented by the reflection of other. By these kinde of glasses or

rather frames of them, placed in due angles, ye may not onely set out the proportion of an whole region, yea represent before your eye the lively ymage of every towne, village, &c. and that in as little or great space or place as ye will prescribe, but also augment and dilate any parcel thereof.[28]

To judge from this brief account, Digges was suggesting that the concave mirror was trained on the object under scrutiny and that the image reflected from it would have subsequently been enlarged by the lens. Here, in other words, the mirror was the objective and the lens was the eyepiece, and the observer, while looking through the latter, would need to bring the image into focus. Either a convex or a concave lens of short focus could have been used.[29] As the emphasis on the "frames" and the "due angles" of this combination perhaps suggests, the observer's own head would have easily blocked the initial beam from the object to the mirror. In this section and in the preface to the reader, Digges attributed the device to his late father and declared that it was suitable for scrutinizing small objects like coins or letters some seven miles off.

This very spare description of the instrument was evidently intended as the selling point of his entire work, the subtitle of which promised *Sundry strange conclusions both by instrument and without, and also by Perspective glasses, so set forth the true description or exact plat of an Whole Region.* Whereas the first edition did not feature a telescopic view on its title page, the revised version of 1591 offered a prominent place to the Euclidean means of measurement with a plane mirror, above which hovers, in a semicircle near the sun, "the lively ymage" of a particular town. The promise that the combination of a concave mirror and glass lens would allow one first to "set out the proportion of an whole region" or to "set forth [its] true description or exact plat" or map, and then to magnify any of its particulars, suggests a progression from a surveyor's schematic rendering of a landscape to a kind of telescopic vision.

Rather than develop the subject further in the *Pantometria,* Digges alluded to a separate volume devoted to "the miraculous effectes of perspective glasses," but no such work appeared. Digges did, however, insist on his father's role again in a military treatise of 1579, and here he sought to give the invention a Baconian pedigree; Leonard Digges had been assisted, he said, "by one olde written booke of . . . Bacons Experiments, that by straunge adventure, or rather Destinie, came into his hands."[30] He went on to explain that though he had planned to spend his time testing out his father's conclusions, constant lawsuits had deterred him. Legal wrangles did, in fact, consume Thomas Digges in those years, and it is curious to note that a century later a member of the Royal Society alleged that there was "a great controversy" between Digges and his chief antagonist, the Kentish Lord Chief Justice Roger Manwood, as to which of them had "revived" the telescope after its original invention by Roger Bacon.[31]

Some sort of telescopic device, in any case, was being revived in this period by yet another Englishman, William Bourne, a mathematical practitioner and councilman of Gravesend.[32] Bourne's writings on the optical instrument are more explicit than those of Dee or Digges, and more persuasively suggest actual experimentation with combinations of lenses and mirrors.

Though his *Inventions or Devices* was published in 1578, it had emerged about two years earlier as a manuscript dedicated to Sir William Cecil, the first Baron Burghley, whose patronage he hoped to obtain. Whereas Digges had presented his lens-and-mirror combination as a corollary to a context reminiscent of the Pharos, progressing from the observation of approaching ships in a seaside tower to a description of a telescopic device, Bourne advanced from the trick of conveying reflections from one mirror to another over a short distance to that of producing magnified images of remote places.

The first device, a periscope-like affair, offers a scaled-down version of the Pharos to the port master or prosperous consumer. Bourne asserted that a series of large looking-glasses placed in "a very high house that hath windowes that are of a great height from the floor or else some high Tower" might show the approach of ships in the harbor or persons on the highway, or, in the case of country gentlemen, the "Deere in their parkes, or cattell in their pastures, or what persons that there is stirring in their Gardens or Orchards." In claiming that such instruments were "very necessary either for men of Honour or Gentlemen," and in restricting their usage to those who happened to have on hand outsize mirrors and to live in lofty buildings, Bourne avoided the suggestion that he wrote for people whose residences were precariously close to war zones or who spent their time spying upon "yonder bedroom," but the device worked no better than that described twenty-five years earlier by Cardano and was vulnerable to the same criticism. Tellingly, a marginal notation in the manuscript version pointed out that "the reflection will be very weak."[33]

Bourne's description of his telescopic device is somewhat more convincing. He recommended the use of a very large convex lens as an objective and a good-sized concave mirror as an eyepiece. In emphasizing the change in appearance of objects perceived through the convex lens—as one moves away from the lens, the images suddenly go from being upright, increasingly enlarged, and ever more blurry to inverted, progressively contracted, and sharp—Bourne sought to represent the point of greatest magnification, for this spot was where the focal point of the concave mirror likewise needed to be. An observer looking into the mirror would recognize a man a mile off, he promised, but he added the crucial caveat that such effects would be achieved only if the convex lens were "well made and very large."[34]

Around 1580 Bourne returned to this device in a brief treatise on lenses and mirrors destined for Lord Burghley. This time, he insisted on the utility of the device to military men, and as before described the changing appearance of images seen through a convex lens, and alluded to the alignment of a very large magnifier with an outsize concave mirror. The point of the treatise seems to have been to obtain Lord Burghley's fiscal support, and when he addressed the specifics of the device, Bourne suggested that the expense of the components had prevented his experimentation with mirrors and lenses of the highest quality. Noting that Thomas Digges and John Dee had access to more leisure, books, and money, he implied that they, rather than he, ought to undertake further research, and that the proper combination of lens and mirror would reveal images "of a marvelous largenesse, in manner uncredable to bee believed of the common people." The treatise ended by insisting on the reliability of Thomas Digges's allusions to his father's accomplishments with the telescopic device.[35]

Dee, Digges, and Bourne all implied that the telescopic device was relatively unknown, and they appear to have worked, to varying degrees, to keep it out of the public view by describing it in vague terms or in texts not intended for publication, by limiting their printed remarks to the obscure English vernacular, and by awarding it a pedigree dependent upon a rare Baconian manuscript.[36] That said, a curious figure tangential to their orbit, the Kentish author Reginald Scot, suggested a certain familiarity with "perspective glasses," with the ongoing discussion of the matter in Italy, and with telescopic vision's debt not to the latecomer Roger Bacon but to Apollonius of Tyana and the Pharos. In his *Discoverie of Witchcraft,* published in 1584 and dedicated to Thomas Digges's archenemy Roger Manwood, Scot held forth about the usual tricks of perspective, among which were "glasses . . . so framed, as therein one may see what others do in places far distant," and "clear Glasses, that make great things seem little; things far off to be at hand."[37] He ended two pages of staccato references to mirrors and lenses with the declaration that "these I have for the most part seen, and have the receipt [recipe] how to make them; which if desire of brevity had not forbidden me, I would here have set down. But I think not but *Pharaohs* Magicians had better experience then I for those and such like devices. And as (*Pomponatius* saith) it is most true, that some for these feats have been accounted Saints, some other Witches."[38]

Scot's statement is important for several reasons. The English author ignored the rationalist thrust of Pomponazzi's argument—the latter had suggested that atmospheric reflection, *not* pharaonic feats of the supernatural sort, had allowed people to see remote events—but he did return the discussion of telescopic vision to the Italian context, and beyond that to Apollonius and the Pharos. And although the conventional excuse that his "desire for brevity" was the sole impediment to further information about these optical devices ought to be treated with suspicion, in view of the considerable length of the *Discoverie,* there is some evidence that Scot had in fact seen either a telescopic instrument or specifics about its construction.

In the early 1580s, as he was composing the work on witchcraft, Scot also observed and wrote of the repairs made to Dover Harbor, an enormous construction project in which Thomas Digges played an important part as surveyor and consultant from 1576 through 1584.[39] Given that Digges had originally presented his father's "perspective glass" in the context of estimating the distance of ships from a castle-like structure such as the one at Dover, and that telescopic images of remote parts of the harbor would have been useful during the protracted process of its reconstruction, it is feasible that in these circumstances he disclosed more about the instrument to Scot. And though the image that accompanied Digges's brief account of the "perspective glass" is neither very realistic nor detailed, it does evoke the crucial features of England's most famous harbor in that period: a castle set high on a cliff, a small bay, and a vast stretch of silt that made the port impracticable.

Scot's allusions to Pomponazzi's discussion of telescopic sight, and his insistence on the antiquity of the device, if prompted by his acquaintance with Thomas Digges, do much to undercut that alternate pedigree provided by the English mathematician, one that made the "perspective glass" the invention of the late Leonard Digges and the brainchild of Friar Bacon. In effect, Scot's account reinstates another old Italian at Dover Castle and ultimately returns the invention to the Pharos of Alexandria.

One other work of this period appears to address the general question of the Baconian origins of Digges's device, though without ever explicitly mentioning the "perspective glass" itself. This was Edward Worsop's *Discoverie of Sundrie Errours and Faults Daily Committed by Land-Meters,* published in London in 1582. Staged as a dialogue between speakers of various backgrounds and social origins, Worsop's text praises Robert Recorde, John Dee, and both Diggeses as geometers, not as inventors, and is forthright in its criticism of surveyors whose showy instruments mismeasure land and defraud the public.[40] In a defense of the specialized Greco-Latin vocabulary then proper to geometers, Worsop explains that such terms are neither "Pedlars' French"—a kind of thieves' argot or canting language—nor a collection of conjurations used to summon evil spirits.[41] But this latter possibility—or rather the suggestion of a strong popular association of geometrical and occult knowledge—is raised by a simple serving man named Steven, and it lingers throughout the dialogue.

Steven, plainly illiterate, relates that he and his friends encountered a book inscribed with crosses, figures, circles, and foreign words, an ambiguous object that may have been either a work of geometry or a *grimoire* filled with magical drawings and incantations, and it is significant that the volume was found in the house of a Catholic priest with remarkable skill in measuring and surveying.[42] Steven's description of Father Morgan's instrument, while presumably the embodiment of naïf reportage and candor, is similarly ambiguous: he recalls that the priest looked through "a fine knacke or jig," "and measured a good pretie way from him," and that after a second sighting, and "casting a little with his pen," the man gave the correct answer.[43] There is no need to assume that either trickery or optical components were involved, but because "knacke" and "jig" meant "contrivance" or "device" or "ruse," and because "casting" indicated both legitimate reckoning and occult forecasting, the priest's activities and instruments are never entirely freed from such connotations.[44]

Insofar as the dialogue goes on to detail the way in which centuries of popery and superstition had corrupted the popular impression of mathematicians, the priest's position looks increasingly suspect; indeed, Steven's innocent allusion to Father Morgan's skill in "undermining" things would have gained new resonance after the Gunpowder Plot of 1605.[45] Though the priest's alarming ability in warfare—his knowledge of the size of an enemy camp, or how best to scale walls or to aim gunnery—is consistent with the skills broadly recommended to all sorts of men, "of what vocation or degree so ever he be" in Thomas and Leonard Digges's *Pantometria,* such abilities recast him, in sum, as an updated and unsavory version of Friar Bacon, albeit one whose instrumentation remains ambiguous.[46] As Chapter 3 will show, there was no end to the odd suggestion that something hidden or occult accompanied surveying and measuring procedures, that mathematical formulae had some kinship with magical formulae, and that the surveyor's "fine jig or knacke" merited special scrutiny: it would reemerge in Galileo's orbit in 1607.

An Instrument for Seeing Far

The Italian interest in telescopic devices grew steadily in the latter decades of the sixteenth century. Ettore Ausonio died around 1570, and his chief successor appears to have been the Neapolitan playwright and magus Giambattista della Porta.[47] The first edition of della Porta's *Natural Magic,* published in Latin in 1558, included a number of optical tricks, but it offered no means of observing remote objects except for the usual periscope-like arrangement of mirrors. Fifty years later, when the Dutch telescope first appeared in The Hague, della Porta was widely identified as its inventor, in large part because of claims made in the second edition of *Natural Magic,* which emerged in 1589 and promised an improvement upon the device once deployed in the Pharos.

Toward the end of 1579, Cardinal Luigi d'Este invited della Porta to conduct his research in his villa in Tivoli, near Rome. Though the cardinal was interested in della Porta's productions as a playwright, his dabbling in alchemy, and his skill as an astrologer, his optical projects would not have been alien to his patron. Ferrara, ruled at that time by the cardinal's brother, Duke Alfonso II d'Este, would see the publication of Rafael Mirami's *Compendious Introduction to the First Part of Catoptrics* in 1582, and the duke himself was the patron of another magus, Abramo Colorni, a Mantuan engineer, architect, and occultist whose instrument, like that of Thomas Digges, was allegedly both a rediscovery of an older device and an outgrowth of a surveying tool. Della Porta must initially have been judged a mere follower of Colorni, for when the latter failed to see his long-promised treatise into print, his close friend Tomaso Garzoni emphasized its primacy in 1585 by referring to those who threatened to outshine the Mantuan. "And who does not know," he wrote in the prefatory letter to Colorni in his *Piazza Universale,* "that in the mathematical disciplines you leave behind you so many imitators, both in the past and at present, having with your lofty spirit discovered anew instruments for measuring with sight that are easier, clearer, more useful, and for farther distances than any others, as you will show in offering your most learned book *Euthimetria* like a lucid mirror to the world?"[48]

Apart from the pressure to succeed where Colorni had not, della Porta might also have conceived of a telescopic device as a particular means of aiding his patron, for the cardinal had suffered from some sort of visual defect due to a severe injury to his left eye in adolescence.[49] After some months near Rome, by late 1580 della Porta was in Venice to oversee a project of special interest to his patron, the manufacture of a large parabolic mirror. An envious contemporary reported that della Porta was trying to make "an instrument for seeing far" and that he refused to show his secret to anyone.[50] Two individuals in whom he evidently did confide were the Venetian senator, patron of the arts, and amateur scientist Giacomo Contarini and the Servite friar, state theologian, and polymath Paolo Sarpi. Contarini, described as "a new Archimedes," was the owner of an important scientific library and instrument collection, and he oversaw the difficult and not entirely successful production of a parabolic mirror in November 1580 at the Arsenal. Sarpi, whom della Porta would praise in his *Natural Magic* of 1589 as "the glory and ornament of neither just the city of Venice, nor of Italy alone, but of the entire earth," likewise engaged in experimentation with lenses and mirrors around 1580, and his research, particularly as it pertains to Galileo, is the subject of the next chapter.[51]

What seems clear about della Porta's work in the 1580s is that he regarded a concave mirror in combination with a glass lens as the means of producing telescopic effects, and that he did so in part because the recent appearance of stories about the Pharos at Alexandria added new authenticity to the familiar romance motifs of the Virgilian and Far Eastern devices. Such views emerge in the 1589 edition of his *Natural Magic,* published shortly after the death of his patron. Consider, in this connection, the proem of book 17 of this work, the subject of which was the use of mirrors and lenses.

One reads that Archimedes of Syracuse vanquished the Roman forces by means of burning mirrors, and that King Ptolemy built a tower in the Pharos in which he placed a mirror, such that he was able to see from a distance of six hundred miles the enemy ships that were invading and plundering his territory. And to this I will join glass lenses by which means dim-sighted men may see all things perfectly, even objects at a great distance. And if it seems that antiquity thought up many and great wonders, what we will deliver is yet greater, more majestic, and more illustrious, and of no small benefit to those who aspire to the science of optics, such that it will be able to flourish infinitely in the highest minds. And finally we will show how to fashion and polish both glass and metal mirrors.[52]

The proem offers the general plan of book 17 of *Natural Magic,* which progresses from mirrors to lenses to the manufacture of the former. It also suggests, in the absence of explicit references to the trick mirrors with which much of the book is mainly concerned, that what most interested della Porta in 1589 was the relationship of looking glasses and lenses to telescopic vision, and this much is confirmed by a letter about the manuscript that he wrote in 1586 to Cardinal d'Este. "I will bring my book on the marvels of nature, which I started more than thirty years ago," he confided, "and in which I have placed every secret chosen and proven by all sciences, that is, the very subtlest things, and the core of each discipline. As, for example, in the case of perspective, the way to make a mirror that burns something a mile away, and another with which one might converse with a friend a thousand miles distant by means of the moon, and how to make glasses [*occhiali*] that show a man a few miles away, and other miraculous matters."[53]

The sorts of mirrors noted here are both concave. The first, of course, was associated with Archimedes' legendary immolation of the Roman fleet at Syracuse, and the second was of the sort popularly believed capable of reflecting a message onto the lunar surface such that it was visible to viewers in remote locations, a typical early sixteenth-century claim being "nothing happened in Milan that was not known in Paris that night." By the early seventeenth century the so-called Pythagorean mirror would be rumored to be systematically employed for communication between Naples and Spain, between Constantinople and England, between Prague and Paris, or, more prosaically, between Paris and London.[54]

The question here is which features, if any, these bogus mirrors shared with the telescopic *occhiali* subsequently mentioned. It is crucial, therefore, that we read della Porta's phrase in the proem, "and I will join to this [concave mirror] glass lenses by which means dim-sighted men may see all things perfectly, even those at a great distance," to describe not just the organization of book 17, nor just the triumphal narrative of modern improvement upon ancient discoveries, but also the actual physical configuration of the two components of an optical device.

This proposed combination of lens and mirror is reiterated in the eleventh chapter of book 17 of *Natural Magic,* and as before it proceeds from an allusion to the Pharos to an obscure description of a telescopic device involving a concave cylindrical mirror and some sort of spectacle glass. This account was so contorted that Johannes Kepler noted, soon after the invention of the Dutch telescope, that della Porta purposely left unclear whether two lenses or a lens and mirror were to be adopted, and other contemporaries alleged that Paolo Sarpi alone could follow it. Entitled "Of lenses, by means of which one may see very far, beyond imagination," della Porta's eleventh chapter promised valuable optical information to an elite few and confusion for the common reader.

> I will not omit a marvelous and most useful thing, whereby dim-sighted men are able to see far beyond what can be believed. I spoke of Ptolemy's mirror, or rather glass, with which he could see ships approaching from six hundred miles away, and I will try to teach in what way it could be brought about that we can recognize friends at several miles' distance, and read the smallest letters,

though barely discernible, from afar. This is a thing necessary to man, and founded on optical principles. And it can be done with a trifling amount of work, but it is not a thing to be spread among commoners, and yet is clear from perspective. Let the strongest [sight ray] be in the center of the mirror, such that all the sun's bright rays are dispersed, and do not converge at all, except for in the middle of the aforesaid mirror. There the transverse rays all cross each other. In this way a concave cylindrical mirror with equidistant sides is made, but let it be joined on one oblique side by these sections. Indeed [these sections] of either obtuse or right-angled triangles should be cut from different directions by two transverse lines, drawn from the center. And a glass lens will be prepared for it. I have said what is necessary.[55]

The passage is a model of obscurity, and a later reader was surely correct when, like Kepler, he suggested that della Porta intended to mystify most of his audience and that some of the unfortunate latter sentences were out of order or otherwise garbled by an editor rather than by the author alone. That reader's inference that the perplexing references to "sections" concern conic sections is also sound: here della Porta meant to indicate hyperbolic and parabolic surfaces, those curves traditionally understood as sections of circular cones whose vertex angles were respectively obtuse or right.[56]

More to our purpose is the suggestion that some sort of concave mirror, used in combination with some sort of lens, would produce a telescopic effect. It is gratifying to find, after the overall obscurity of his account of the device, a bit of enlightenment about the relevant lens in a passage from one of della Porta's comedies. The term he used to describe those who would benefit from his invention, *lusciosi* or "dim-sighted," is one he also adopted in the proem where he proposed to "join glass lenses by which means dim-sighted men may see all things perfectly, even those at a great distance." Though in antiquity the word had indicated either those who suffered from night blindness, a condition generally associated with myopia, or from myopia alone, by the early modern period it and its Italian equivalent referred more ambiguously to people with any sort of defective sight, to which category his late patron the cardinal would have belonged. The most celebrated usage of the term in classical literature, in any case, occurred in Plautus's *Braggart Soldier,* a play so well known to della Porta that one of his own comedies was judged to be nothing but a good translation of the Latin original.[57] In the Plautine context, when two servants are debating whether or not one of them actually saw adulterous activity in the house next door—he did—the insult of choice is *luscitiosus,* and because the events take place by day, myopia rather than night blindness is the relevant condition.[58]

In della Porta's farce *La Chiappinaria,* the context itself recalls both the Plautine model and Girolamo Cardano's pairing of sights of "yonder bedroom" with those of nearby battlefields: Cogliandro, a foolish and overprotective father, summoned by his neighbor to witness "great marvels," anticipates seeing "combatants in a stockade, barricaded armies, massacres of men," and instead observes the daytime escapades of his own daughter and her suitor. She seeks to persuade her father that his eyes have fooled him, and in this case the dim-sighted dupe is quite clearly one who suffers from myopia as well as from gullibility:

Cogliandro: The Captain called me up onto the roof of his house, and from there I saw you with Albinio in your bedroom.
Drusilla: Did you see this, sir, with your glasses, or without?
Cogliandro: With glasses, the ones I use when I want to see better.
Drusilla: Maybe they were those glasses that distort, and make you see one thing for something else.[59]

If we assume then that the *lusciosi* mentioned in *Natural Magic* are synonymous with myopes, the configuration della Porta had in mind for telescopic vision would have had a concave mirror as the objective and a concave lens as an eyepiece. This is perhaps the arrangement described in 1571 by Thomas Digges in the *Pantometria,* and in 1632 Galileo's student Bonaventura Cavalieri would note in his work on burning mirrors that such a combination would produce a telescopic effect and that this configuration might have been "Ptolemy's mirror" in Alexandria.[60]

Della Porta's vagueness about both the mirror and the lens, in any case, would have been motivated by an evident desire for secrecy about what he recognized as a still imperfect but potentially valuable invention. His scholarly treatise on refraction, published in Latin in 1593, included various discussions of topics related to both magnification and the observation of distant objects: della Porta alluded to the Senecan trick of enlarging fine print by viewing it through water-filled glass globes, he attributed the apparent increase in size of celestial objects near the horizon to atmospheric refraction, he noted that crystal spheres and convex lenses magnified nearby things, and he observed that concave lenses produced sharper images for the dim-sighted. He did not, however, mention telescopic devices of any sort, even while discussing historical estimates of the limits of human sight, and he repeated the apocryphal story of a man who could count ships in the port of Carthage from a perch in Sicily without ever referring to the optical adventures of "King Ptolemy" in the Pharos.[61] Such restraint suggests that the telescopic device mentioned in *Natural Magic* was still at best a rather flawed instrument, but it is also clear that della Porta continued to experiment with lenses and mirrors, demonstrating their uses to other students of natural philosophy in 1604 and even in the fall of 1608, shortly after the invention of the Dutch telescope.[62]

A World of Glass

Given its vagueness, della Porta's description of the instrument in the Pharos might also have been associated with the camera obscura, for chapter 6 of book 17 of *Natural Magic* describes that device in much more explicit fashion. Della Porta's innovation was to use both a convex lens in the aperture of the camera obscura and a concave mirror at a suitable distance from this opening; the upright and somewhat enlarged image would have subsequently been reflected from the mirror onto a sheet of white paper. Some readers would perhaps have recognized a family likeness between this arrangement and the telescopic device described by Bourne in 1578 and 1580: there, a convex lens was trained on the remote object and viewed in a concave mirror. Given the legibility of della Porta's account of the camera obscura and the studied imprecision of his depiction of the instrument used by "King Ptolemy," in other words, one could easily infer that the Pharos was a dark room equipped with a convex lens and concave mirror.

Such an assumption, in any event, might well have been operative for the first readers of Edmund Spenser's allusion to a telescopic mirror of some sort in the first English epic, the *Faerie Queene,* published in 1590, one year after the enlarged second edition of the *Natural Magic.* At stake is a concave spherical looking glass devised by Merlin, an occasional English substitute for Virgil the necromancer.

> It vertue had, to shew in perfect sight,
> What ever thing was in the world contaynd,
> Betwixt the lowest earth and heavens hight,
> So that it to the looker appertaynd,
> What ever foe had wrought, or frend had faynd,
> Therein discovered was, ne ought mote pas,
> Ne ought in secret from the same remaynd;
> For thy it round and hollow shaped was,
> Like to the world it selfe, and seem'd a world of glas.[63]

Merlin's glass recalls its Chaucerian forebear in "The Squire's Tale." Like the earlier poet, Spenser deployed the imperial mirror in an amorous context, describing the exotic counterpart to his Arthurian device as a prop in a love story, albeit one that perishes in an absurd quarrel in the dark chamber where it was once deployed. Unlike Chaucer, however, Spenser made no mention of the naturalness or feasibility of the instrument, treating it instead as the occult emblem of an irretrievable epoch:

Who wonders not, that reades so wonderous worke?
But who does wonder, that has red the Towre,
Wherein th'Egyptian *Phæo* long did lurke
From all mens vew, that none might her discoure,
Yet she might all men vew out of her bowre?
Great *Ptolomæe* it for his lemans sake
Ybuilded all of glasse, by Magicke power,
And also it impregnable did make;
Yet when his love was false, he with a peaze it brake.[64]

To conclude, what is especially interesting about Spenser's account of the looking glass is the way in which its relentless bid for antiquity and a kind of documented inauthenticity would have been undercut by historical circumstance. At least some of Spenser's earliest audience, newly acquainted with the story about the Pharos and aware of it not as a glass tower demolished by jealous Ptolemy but rather as a still-visible ruin, would have also encountered devices and inventions like those proposed by Digges, Bourne, and della Porta. Far from assuming that the instruments in the *Faerie Queene* were confined to Merlin's grotto or had met an ignoble end in a domestic dispute in ancient Alexandria, some would have believed that they worked, sometimes and somewhere.

———⇒·◆·⇐———

Chapter Three

Obscure Procedures and Odd Opponents

𝒜NY attempt to gauge the reactions of Galileo and his close associate Paolo Sarpi to the rumor of the Dutch telescope must naturally begin with an assessment of their understanding of telescopic vision prior to their encounter with the news from The Hague. Sarpi's optical study from the late 1570s through 1600 and Galileo's interest in the discipline from his arrival in the Veneto in 1592 through 1607 are of particular importance. Although it would be unwise to assume that these men held identical views about the production of telescopic effects, both were familiar with the ongoing discussion about concave mirrors, and both were acquainted with improvements to the camera obscura and exercises in altimetry. The means of measuring the height or distance of remote structures were at this point incompatible with telescopic effects, but some part of the interest in Galileo's instruments derived from a mistaken belief that such procedures might be combined in a single device. Those expectations—which were clearly an outgrowth of those attendant on

Abramo Colorni's undelivered promise—provide the best index of a discussion that died away relatively soon after the invention of the Dutch telescope.

Reading Letters Fifty Paces Away

Fra Paolo Sarpi's interest in optics can be partially reconstructed through his notebooks, his decades of correspondence, and the accounts of his contemporaries. Each sort of evidence poses interpretative problems: Sarpi's annotations are sometimes undated and illegible, his terse and cryptic letters were so detached in tone that their authenticity and provenance were occasionally questioned even by those who intercepted them, and older biographical information about his unnoticed discoveries and inventions verges on the hyperbolic. What is striking about the attention Sarpi gave to optics, particularly to actual devices, is its close relationship to contemporaneous discussions of the discipline in northern Italy in the late sixteenth and early seventeenth centuries. In this regard, he appears a representative figure.

Sarpi had already been investigating mirrors for at least two years when he encountered Giambattista della Porta in 1580 during the latter's attempt to obtain a large concave mirror and a means of making an instrument "for seeing far." Sometime during the three years in which he was in Rome—1585 through 1588—Sarpi made a brief trip to Naples to visit della Porta, who was then engaged in finishing the 1589 edition of *Natural Magic.* He also undertook a study of the human eye itself in collaboration with the celebrated anatomist Girolamo Fabrici d'Acquapendente of the University of Padua. It was probably around 1592 in the home of Padua's best-established intellectual, Gianvincenzo Pinelli, that Sarpi met Galileo, newly arrived in Venice, as well as the Venetian Gianfrancesco Sagredo, an enthusiastic student of optics, magnetism, tides, and mechanics. Though the theologian's notations about optics and most other subjects of natural philosophy appear to have tapered off toward the end of the century, Sarpi subsequently took up such issues, particularly if they figured as news or novel developments, in his correspondence.[1]

Sarpi's notebooks reveal his engagement with optical topics typical of the later sixteenth century, some of which were relevant to telescopic vision. He was interested, for instance, in the comparison of plane, convex, and concave mirrors, and in the use of the last in projecting light for reading or for scrutinizing others at night, and in propagating sound or heat, all issues previously examined by Ausonio, Cardano, and della Porta, among others. His attention to the chemical composition and decomposition of glass, and its comparison with crystal, while hardly surprising in a Venetian setting, is also characteristic of those who sought out the best lenses and mirrors.[2]

Here and there in his notes Sarpi focused on the ways vision might be enhanced. While acknowledging that glass and other transparent media could alter the apparent size of a nearby object, Sarpi seemed unwilling to accept the traditional argument about the role of atmospheric vapor in the magnification of celestial bodies on or near the horizon, suggesting instead that the observer's estimate of the size of such bodies was conditioned by his illusory impression of the distances involved. Sarpi's meditation on the problem of an observer's reckoning of size and distance is especially interesting in terms of a crude telescopic effect much discussed in the wake of the Dutch invention. Arguing that the observer tended, under normal circumstances, to judge an object's remoteness by comparing it with other things closer to him, he noted that these relative estimates failed in the case of great distance, producing an illusion that makes "the stars look like they are all on one surface, or roads and mountains appear to be touching the sky, or a somewhat distant light surrounded by fog or by night seem like a star." In other words, a kind of misapprehension made more remote objects appear either somewhat closer, as in the case of distant stars and the horizon, or slightly bigger and brighter, as in the case of the starlike light wrapped in fog or darkness. Sarpi ended this discussion by noting that the same effect was produced when one looked through a long tube known as a *cerbottana,* a primitive blowgun adapted, in this case, for optical activity.[3]

References to what could be seen through the cerbottana, through one's clenched fist, and from the bottom of a well were common in this period, and prior to the invention of the Dutch device these practices were regarded as primitive and related means of achieving telescopic vision.[4] Such statements ultimately derived from the fifth book of Aristotle's *On the Generation of Animals*—the probable source for Sarpi's confusion of presbyopia with glaucoma, and for his proper understanding of the notable differences in vision among various species—where the ancient philosopher distinguished between the ability to see distant things and the ability to bring them into focus.[5] Noting that an individual might not have both capabilities, Aristotle added remarks that would occur again and again in discussions of both the Dutch telescope and its precursors:

> The man who shades his eye with his hand or looks through a tube will not distinguish any more or any less the differences of colours, but he will see further; at any rate, people in pits and wells sometimes see the stars . . . Distant objects would be seen best of all if there were a sort of continuous tube extending straight from the sight to that which is seen, for then the movement which proceeds from the visible objects would not get dissipated; failing that, the further the tube extends, the greater is bound to be the accuracy with which distant objects are seen.[6]

Such observations offered the basis for the loose association of the cerbottana with telescopic vision in the aftermath of the invention. A friend of both Galileo and Sarpi, Raffaello Gualterotti, for instance, contested the glory of the anonymous Dutch inventor, in part because he felt that such credit naturally belonged to a Florentine, but also because he saw both his own use of the cerbottana as a means of observing stars by day in 1605 and his alleged development of another optical instrument "for jousting and for war" around 1598 as worthy precursors of the device from the Netherlands. Even though Gualterotti himself called that invention "a puny thing," he associated it and the cerbottana with his early experimentation, around 1580, with a pinhole aperture in a camera obscura.[7] Gualterotti's guide in this experiment was almost certainly the first version of della Porta's *Natural Magic*, where the reader was invited to descend into "the deepest of wells" in order to see stars by day, and in the following chapter offered, as if by default, a much more manageable procedure, that of making a pinhole aperture in a dark room.[8] The fact that the pits of wells, the cerbottana, the camera obscura without a lens, and the eventual telescope all involved restricted visual rays and diminished circumambient light, in other words, persuaded some to see these earlier devices as clear antecedents of the Dutch invention.

But Sarpi's concerns with enhanced vision, fortunately, were not limited to the modest effects brought about by cumbersome tubes or lonely vigils in the depths of wells. To the extent that one can judge from a brief series of notations made between 1578 and 1583—in the period of his initial acquaintance with della Porta—he associated telescopic sight with concave mirrors, almost certainly in combination with lenses. The general context in which these instruments would have been conceptualized, if not developed, is at once that of measuring heights or distances, as was the case for Thomas Digges and Abramo Colorni, and that of the camera obscura, as was the case with William Bourne and della Porta.

Measure the height of something with the sun's shadow, with the shadow of something else, and with the mirror. ▲ Writing that has been reversed can be read very easily in a mirror. ▲ One or more mirrors can be positioned so that a man can see whatever is done outside, and likewise by arranging glasses. ▲ So that letters can be read from fifty paces [that is, ninety-four yards] away: I tried it with the spherical [mirror], and / or with the lens, but it is better with the parabolic one and / or with its lens, and reading them with the light source far away.[9]

Chief among the difficulties of this passage—as the awkward "and / or" shows—is the fact that Sarpi's handwriting leaves unclear whether he intended to read letters from fifty paces by combining a concave mirror and a lens, or by substituting one optical component for the other. The wording in his allusion to the camera obscura as a covert means of surveying activity outside is likewise ambiguous: it is not initially evident whether Sarpi intended to adopt both a convex lens in the aperture and a concave mirror within, as della Porta would in *Natural Magic,* or was satisfied with either the inverted image produced by a mirror alone or with the reversed one offered by a convex lens.

What does in any case seem likely is that the telescopic effect described in the last annotation was understood to occur within the dark room, for users of the camera obscura sometimes stipulated that the illumination necessary for the procedure had to be on the object under scrutiny but not near the aperture.[10] Furthermore, in order to read writing from within the dark room, whatever its distance, a lens *and* a mirror would be crucial to producing an upright and unreversed image: all other texts would be useless to the observer. For this reason, the brief second annotation also makes much better sense if one assumes that it, too, involves the camera obscura, for the circumstances in which writing would appear reversed, but not inverted, would be when it had passed through the convex lens in the aperture of a dark room. Once reflected by the concave mirror, the text would be perfectly legible.

It appears, therefore, that Sarpi was familiar with a dark room setup like the one described a decade or less later by della Porta in book 17 of *Natural Magic,* and that it worked well enough to render writing some fifty paces away legible for an observer within the camera obscura. The several ambiguities of Sarpi's annotations bear a certain resemblance to those surrounding della Porta's work. As suggested in Chapter 2, della Porta's work on optical devices around 1580 was probably seen as an alternative to something telescopic promised by Abramo Colorni and associated with measuring the heights of remote objects; in the words of Tomaso Garzoni, once they were disclosed in a book compared to a "lucid mirror," Colorni's newly "rediscovered" instruments would offer ways of "measuring with sight that [were] easier, clearer, more useful, and for farther distances than any others." Colorni's devices, never realized, would have been an elaboration, or rather a combination, of the kinds of procedures outlined by Thomas Digges, which progressed from the use of a plane mirror to estimate distances to the deployment of a concave mirror as a telescopic component. The mirror would have served as the objective, and a glass lens as an eyepiece, and the whole procedure would have taken place outside. Della Porta's ambiguous proposal to replicate the instrument at the Pharos by joining to a concave mirror the sort of glass lens favored by "dim-sighted men" could well have involved this configuration.

At the same time, the vagueness of della Porta's account and the relative lucidity of his description of the camera obscura had made it possible for readers to understand the latter device as the telescopic one. In this case, the convex lens in the aperture naturally served as the objective, and the observer in the dark room would look for the telescopic image either in the concave mirror or on the paper onto which that image was projected. Put differently, one might find in book 17 of *Natural Magic* two "instruments for seeing far," both involving a lens-and-mirror combination, with no clear indication of which worked better. The fact that Sarpi's first annotation includes a quick reference to the use of a mirror in altimetry, and then passes without explanation to the camera obscura and eventually to a few terse references to its use for telescopic vision suggests that he, too, had both of these alternatives under consideration in this period.

Sarpi's close and reasonably well-documented relationship with della Porta makes it likely that his conjectures about telescopic devices developed within the context of this friendship. That said, Sarpi had some familiarity with Ettore Ausonio's work on concave mirrors, though his copy of the manuscript, made no earlier than 1587 and perhaps later, appears a crude approximation of the original. And although there is a certain similarity between the elements of Sarpi's annotation and adjacent items listed in the schematic section of Ausonio's *Theoretical Discourse on the Concave Spherical Mirror*—viewing sunlit scenes or reading remote letters by night within a camera obscura—the older manuscript appears to make no explicit mention of the crucial role of lenses, despite the fact that these, in combination with mirrors, would have both offered a clearer image and, in the case of the writing, corrected its inversion.[11]

It is also possible that Sarpi knew something of the instruments promised, but never delivered, by Abramo Colorni: Sarpi is said to have learned Hebrew while in Mantua in the early 1570s because of his friendships with members of the city's Jewish community, and he and Colorni were both in the orbit of

Duke Guglielmo Gonzaga in this period.[12] But this conjectural acquaintance would not, in any case, substantially alter what Sarpi knew through his contact with della Porta, for Ausonio's manuscript appears also to have been familiar to the Neapolitan philosopher,[13] and Colorni's work was far richer in the impressions it generated about rediscovered instruments, and in the possibility of combining altimetric and telescopic features in a single device, than in anything it actually delivered.

Galileo as Prometheus

Galileo's knowledge of optics prior to his encounter with the Dutch telescope has been the subject of scrutiny in recent years, and an earlier impression of him as an outsider to that discipline has been substantially revised.[14] To some extent, the sense that Galileo's background had little to do with optics derives from the seeming tardiness of his acquaintance with the news from The Hague, and from his own rather uninformative remarks, in later years, about the device itself. Any attempt to revise the traditional account of his belated awareness of this rumor must, therefore, begin by reconstructing his earlier studies and interests.

Though a meeting between Sarpi, Galileo, and della Porta appears to have taken place in Padua in 1593, it is not evident that their discussions had anything to do with optical devices.[15] What is known is that sometime between his arrival in 1592 and the dispersal of Gianvincenzo Pinelli's library in 1601, Galileo made a careful copy of Ettore Ausonio's manuscript on concave mirrors and would have seen that this work, in contrast to a long tradition, located the point of combustion in a mirror near the place where optical images were most enlarged and most blurred. In other words, Galileo would have been aware that Ausonio's *Theoretical Discourse on the Concave Mirror* articulated what the later expression "focal point" makes explicit. He would have likewise been familiar with the claims made in that work about the ability of mirrors to propagate heat, sound, and light, to render echoes and a variety of images, and to make certain things visible within the camera obscura.[16]

Galileo must have been regarded as an authority on some aspects of optics as early as 1593, for in that year the mathematician Guidobaldo del Monte sought out his opinion on the manuscript version of his *Six Books on Perspective,* which would eventually be published in 1600. Though in early 1593 Galileo served as an informal consultant to Giacomo Contarini, the Venetian nobleman who had helped della Porta in his quest for a large parabolic mirror in 1580, what we know of their discussion concerns the optimal position of oars in galleys and has thus to do with mechanics and the strength of materials.[17]

In this phase of his professional life, Galileo appears to have devoted most of his energy to teaching Euclid's *Elements,* spherical astronomy, and mechanics at the University of Padua, and to offering private lessons in the use of a measuring and calculating instrument and on the principles of military architecture to students in his home, many of whom also boarded with him.[18] His university teaching had no necessary connection with optics, though the *Elements* form the conceptual basis of Ptolemy's *Optics* and the prestige of Euclid's best-known work did much to enhance the optical and catoptrical works attributed to him. Galileo's private tutorials likewise appear to have had no particular emphasis on optics, because the greater part of these lessons involved tuition in "the use of the instrument," but his entries in an account book for the year 1601 show that he taught the rudiments of optics at least three times between June and November to foreign students, in conjunction with their study of fortification.[19]

To judge from the biographical account of Vincenzio Viviani, Galileo's last student, written in the 1650s, the lessons on optics were, like those concerning fortification, spherical astronomy, and mechanics, written up as manuals, and were practical rather than purely theoretical in orientation. Most crucially, they followed upon *gnomonica,* a term associated with sundials but used more generally to indicate measurements involving the shadow cast by a gnomon or stake of a known height in the sight lines of an observer facing an object of unknown dimensions. Though Viviani offers no further details and makes no specific allusion to catoptrics, it is worth noting both that his references to the other treatises are accurate and, more to the point, that this brief description of the inextant treatise "on *gnomonica* and practical perspective" recalls that general progression—likewise associated with Thomas Digges, Abramo Colorni, and Paolo Sarpi—from measurements made with shadows to those involving a mirror, and finally to some form of telescopic vision.[20]

It is possible that some of Galileo's well-established interest in the parabolic path of projectiles had also to do with the parabolic mirror. In late 1597, Thomas Seget, a Scottish student at the University of Padua, was introduced into the circle of the learned Gianvincenzo Pinelli, and upon this patron's death in August 1601 he was briefly appointed administrator of his renowned library.[21] In 1599, Sarpi, Jacques Badovere, and Galileo signed Seget's *album amicorum,* a large notebook in which, following a custom of the day, friends and mentors left emblematic drawings and texts celebrating their acquaintances' merits.[22] The astronomer's inscription is based on the relationship between the rhetorical sense of the word *parable* (a comparison or allegory) and the geometrical figure of the parabola—"Thomas Seget, keep this always as a sign of my friendship and regard for you, just as your virtue is stamped with an indelible mark on my heart"—and is signed, "Galileo Galilei, Florentine nobleman and professor at the University of Padua, in Murano, August 13, 1599."[23] Accompanied by a small drawing of a parabola, such a commemoration implies, at the very least, that the context for their acquaintance was the study of conic sections, the knowledge of which was often applied to catoptrics. Although we cannot know whether Seget shared an interest in mirrors of any sort, it is noteworthy that two other men who signed his album in the Veneto in 1599, Filippo Pigafetta and Marino Ghetaldi, had a demonstrated interest in the burning mirrors of antiquity.[24]

One of Galileo's conflicts with rivals from his Paduan days also suggests, as does his acquisition of Ausonio's manuscript, that his interest in optics had specifically to do with some sort of concave mirror. In September 1602, Galileo received a lively letter from Paolo Pozzobonelli, a former private pupil and boarder and an enthusiastic student of mathematics and alchemy. The letter began by congratulating Galileo on the success of some recent and unknown device of his, and by thanking him for a box of eyeglasses he sent for the use of Paolo's relatives. Most of what Paolo had to say concerned his current inability to take up his mathematical instruments or to return to his alchemical studies, but the letter closes with an interesting postscript about a mutual enemy. "I almost forgot the best part," Paolo related. "That illustrious gentleman who made such a liar out of Ingegneri is going to have to be a big man indeed, because he will make liars out of more geniuses than just Ingegneri alone. His glorious fame has flown here [to Savona], and his feats are stupendous. The one where the thinnest iron armor withstands shots from muskets, even from the large kind mounted on trestles, is the least of them."[25]

The "illustrious gentleman" was Giacomo Antonio Gromo, an elderly alchemist, former soldier, physician, world traveler, and fabulist who claimed to have secret means of concocting poison gases, hurling projectiles, protecting troops and their weapons with ointments and potions, preparing artificial food and wine for soldiers at sea, healing wounds and contagious diseases, making glass and enamel, writing in code, and of course making gold. Curiously, Gromo was in England around 1568–1572, apart from a brief stint fortifying the port in La Rochelle, France, with the Venetian Scipio Vergano in 1569. By 1572 he had incurred the enmity of Digges and Dee's patron Baron Burghley for treasonous impostures. After decades in Padua, Gromo had just accepted a lucrative position at the court of Carlo Emanuele I of Savoy, the son of Ettore Ausonio's patron. The poet and playwright Angelo Ingegneri, having just published *Argonautica,* a poem celebrating alchemy and Gromo's singular abilities, was likewise rewarded with a handsome secretarial position at this court, to the amazement of the Venetian ambassador resident in Savoy. Pozzobonelli's remarks about Ingegneri's role as liar were prescient, for within five years that resourceful poet prefaced his work with a plaintive denunciation of alchemy—perhaps, as he later told a friend, "as a joke"—and simply published *Argonautica* again.[26] Although it is surprising that this rather poor work found its way into print not once but twice in the course of a few years, the most interesting aspect is what it reveals about the enmity between Gromo and Galileo.

After a heroic portrait of Gromo and his achievements, *Argonautica* offers, in a style so bombastic as to verge on parody, brief vignettes of other local celebrities in his orbit. Ingegneri presented these individuals as smiling participants in Gromo's glory, but there is already a hint of rivalry in the verses devoted to Galileo:

> But from the Arno emerges a man of sublime mind.
> It is his judgment and profound knowledge
> That render him most illustrious.
> He takes his doubly celebrated name
> From famous and sacred Galilee.
> Though one could call him a new Euclid,
> This man who turns his face to the Sun,
> And holds in his hand the lighted theft
> Has made a worthy and rare offering
> Of the Promethean kind.[27]

Though Ingegneri's verses, even in the original, have little to recommend them, it is noteworthy that they portray Galileo first in terms of Euclid and secondly as Prometheus. Discussion of the myth of Prometheus was varied, but early modern authors often supposed that the legendary figure had used optical means to steal fire from the heavens. The English physician and poet Raphael Thorius, strongly influenced by the work of Girolamo Fracastoro and by Constantijn Huygens, one of the most enthusiastic users of the Dutch telescope, suggested in the early seventeenth century that the celebrated theft had required a convex lens. Several decades earlier, however, Tomaso Garzoni had reported that Prometheus was commonly regarded as the inventor of the burning mirror, and as late as the 1640s the Jesuit natural philosopher Athanasius Kircher likewise insisted that Prometheus had managed without a flint or witches' incantations, but by exposing "a small portion of the clearest mirror to the sun."[28] If a burning mirror is what Ingegneri intended in his insistence on Galileo's Promethean gesture—and it is difficult to imagine any alternative reading—the rather generic reference to him as "a new Euclid" would likely involve less his abilities as a teacher of the *Elements* than his presumed familiarity with the *Optics* or *Catoptrics.*

The perception of Galileo as a figure capable of stealing fire from the heavens in 1601 complements his documented interest in Ausonio's work on concave mirrors in this period, and perhaps his private lessons on "practical perspective" as well. The feasibility of the burning mirror was, moreover, the subject of some scrutiny in northern Italy in these years. Though Filippo Pigafetta, for instance, had added a cautious note in 1581 by observing that his sources related that Archimedes burned the fleet by using large concave mirrors "as if by a miracle," by 1600 he was overseeing a painting of the exploit for the Grand Duke of Tuscany.[29]

Others stuck to supernatural or occult explanations: in the mid-1580s Tomaso Garzoni took up the bland and ambiguous refrain "as if by miracle" in his best-selling *Piazza Universale,* and in 1602 Bartolomeo Crescenzio, an engineer for the Papal States and a close friend of della Porta, reported that because neither the Archimedean screw nor the burning mirror had ever been successfully imitated, many assumed that they had required demonic aid.[30]

Though Ingegneri's work includes a brief and gratuitous reference to the Pharos, it does not offer any further information about Galileo's activities.[31] It is very likely, however, that Ingegneri knew something of the ongoing discussion of optical issues, particularly as they related to della Porta. Ingegneri is best known today for having served as a close friend, a scribe, and an editor of sorts of the great poet Torquato Tasso—like della Porta, a protégé of Cardinal Luigi d'Este—and his stay at the court at Ferrara coincides with both Abramo Colorni's presence there and the Neapolitan philosopher's entry into the cardinal's orbit in Tivoli. As an aspiring playwright and specialist in stage design, moreover, Ingegneri would have been familiar with the several references to theatrical illumination and optical illusions in book 17 of della Porta's *Natural Magic.* Tasso was a correspondent of both Giovanni Antonio Pisani, the most prominent physician in Naples and a mentor of della Porta, and his son Ottavio Pisani, the dedicatee of della Porta's treatise on refraction of 1593; the poet saw both father and son during his visits to Naples in 1588 and 1592, and it is likely that Ingegneri encountered them during his trips to that city from Rome in 1593. Finally, because of their mutual interest in Tasso and in stage production, Ingegneri had long been acquainted with Filippo Pigafetta, who took up the issue of burning mirrors in Tuscany.[32]

An Ancient Adversary

There is much to confirm Pozzobonelli's impression of Ingegneri as "such a liar" about Gromo's doings; however, the poet's portrait of Galileo as an emergent rival is perfectly accurate. Galileo's livelihood as a university professor in those years was greatly supplemented by the lodging and private lessons he gave to students, and Gromo's claims and seniority had the potential to compromise this business. Both Gromo and Galileo professed expertise in military matters, and though there were enormous, even comic, differences in their approaches, they both devoted great attention to projectiles. Galileo's treatise taught his private students, among other things, how to use his geometric and military compass to calculate the distances of remote targets, adjust the angle of the cannon, calibrate the weight of cannonballs of different dimensions and materials, and convert one quantity to another so as to determine the proper powder charge and to avoid wasting the projectile or exploding the cannon itself. Gromo's manuscript, by contrast, offers more than seventy exuberant pages of "secrets" for making projectiles of various substances and performing a different sort of conversion, that of making them into grenades by sectioning them and wrapping them in rags soaked in naphtha, or filling them with poisonous gases. Galileo's manual on the use of his compass is characterized by quantification, a tendency to abstraction, and a certain quite reasonable fear—he notes, for instance, that a modification he has made allows gunners to adjust the angle of their cannons without approaching its mouth—whereas Gromo's work, being a collection of alchemical "secrets," implies that his hearty disciples have access to dozens of toxic substances and putrefying body parts, no particular qualms about handling or ingesting them, and considerable faith in the antidotes and protective potions whose preparation he also described.[33] Though many of Gromo's "secrets" seem culled from della Porta's *Natural Magic,* neither the context in which these techniques were presented—an unpublished autobiographical epic called the *Gromida*—nor Gromo's febrile allusions to his visions and to his occult activity would have inspired much but amused contempt in Galileo.

This is not to say, however, that Galileo and Gromo served two distinct clienteles, for they shared a patron of sorts, the Venetian patrician Giacomo Alvise Cornaro, and at least one student, Baldessar Capra, the son of Gromo's closest ally, Aurelio Capra. Cornaro, portrayed by Ingegneri in *Argonautica* as a heroic successor to Archimedes for having discovered a new and less costly way of manufacturing gunpowder, had been Gromo's ally at least since 1575, when they both approached Duke Emmanuele Filiberto of Savoy with the announcement that they had discovered the true military order favored by the ancient Romans and would be happy to impart it. At least some of the proposals of Cornaro and Gromo—one for a raft mounted with cannonry for fighting along the Danube, another for a handheld device for throwing grenades from trenches—found favor in 1594–1596 with

Filippo Pigafetta, who said that they had already been tried out in the Venetian Arsenal, and that they were effective, cheap, and guarded with great secrecy. And Pigafetta, who would oversee the depiction of burning mirrors and other ancient devices the following year in Florence, claimed that he actually preferred Cornaro's plan of establishing a modern academy of military science in the Veneto to what he regarded as the Grand Duke of Tuscany's antiquarian endeavor.[34]

But Cornaro was also Galileo's closest neighbor in Padua and on excellent terms with him, and it was in his house that Galileo showed Baldessar Capra how to use the geometrical compass around 1602.[35] Galileo evidently regarded both Aurelio and Baldessar with something between scorn and skepticism, primarily because of their association with Gromo, which came soon after their arrival in Padua in 1597. In a letter written in 1604 to the ruler of Mantua, Duke Vincenzo Gonzaga, he noted that some viewed their "secrets" as very suspect.

> In his early days in this city [Aurelio Capra] earned a living by giving lessons in fencing, until he made friends with the most illustrious Signor Giacomo Alvise Cornaro and with Signor Gromo, from whom he learned some medical secrets. At present he maintains himself by practicing a bit in that field, and certain people hold him in some esteem. But more than a few say that since he has had in recent years the closest friendship with Gromo, that he has gotten from him most of his secrets and certainly the most important ones, if not all of them. And there is no shortage of those who believe that he possesses and at present is exerting himself on the "great work," as they call it.[36]

The continuum of the Capras's interests is worth some scrutiny. Galileo appeared initially to suggest that the elder Capra had once earned an honest, if precarious, living as a fencing master before slouching toward medical quackery and alchemy. However, it is more likely that he was exploiting a perceived connection between the "geometry of fencing" and the occult, as if the man and his son had always been disposed to dabbling in the supernatural. This association, based on the strong resemblance between the obscure tenor and Pythagorean diagrams that characterized both fencing treatises and occult manuals, was at once the sort of thing that was praised and paid for by rich young men eager to learn to defend themselves with the sword through study of the "mysterious circle" and mocked in picaresque novels such as Francisco de Quevedo's *Life of the Swindler,* where an inept fencing master is mistaken for a magician.[37] It is not that Galileo had the slightest belief in the efficacy of the occult, but rather that he viewed both father and son, particularly in association with Gromo, as drawn to a fraudulent display of geometry and given over to a swashbuckling manner largely at odds with his own prudent approach to military matters.

The hint of antipathy and distrust, as well as the specific suggestions about the showy misuse of geometry and the attraction for the occult, would emerge undisguised in Galileo's subsequent encounters with Baldessar Capra. This young man, with the help of his tutor in mathematics, the German Simon Mayr, plagiarized numerous pages of Giovanni Antonio Magini's manual on measuring and almost all of Galileo's recent treatise on the geometric and military compass, and published a Latin version of these works, replete with errors, as his own in 1607.[38] The resultant quarrel with Capra is significant in that, like those lines in Ingegneri's *Argonautica,* it offers several suggestions that others attributed optical expertise to Galileo, this time within the context of a surveying instrument.

Galileo had begun to develop his version of this instrument around 1596, refining it over the course of a decade. The system for measuring the height or distance of remote objects was based on an older right-angled sighting tool, the shadow square, which in turn represented an advance over the original system of measuring the shadow cast by an object or by a gnomon of a known height. In the shadow square, notional shadows—the *umbra recta* and *umbra versa*—were inscribed in twelve increments each on the instrument and, when cut by a plumb line, corresponded to particular angles of elevation. This procedure allowed the observer to construct similar triangles to calculate the distance or height of the object under scrutiny. Galileo's innovations were first to transfer the process from a rigid square to his hinged compass or sector, having the quadrant arc show the angle under consideration, and eventually to simplify calculations involving sines or cotangents by discarding the arc and the degree measurements, and by marking the outermost edges of the device with a scale named "Arithmetic Lines" and indicating distance. This latter change was made in 1598 and 1599.[39]

Galileo's adaptations of the shadow square and the sector made it much easier to find the height or distance of a remote object, but the process itself was increasingly removed from the origi-

Altimetry with a shadow square; Niccolò Tartaglia, *Quesiti et inventioni diverse* (Venice: Nicolo de Bascarini, 1554), 25. Rare Books Division, Department of Rare Books and Special Collections, Princeton University Library.

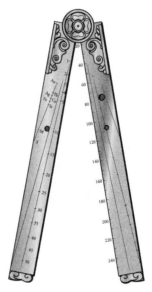

Galileo's geometrical and military compass, ca. 1599. Line drawing by Margaret Nelson, adapted from Stillman Drake and Guglielmo Righini.

nal context of altimetry, where calculations were based on the observer's measurement of shadows cast by the object itself, or by a gnomon of known height that acted as a literal "stand in" for a tower beyond it. The overall progression, in other words, was like that suggested around 1580 in the first part of Sarpi's annotation—"Measure the height of something with the sun's shadow, [or] with the shadow of something else"—but where Galileo turned to an instrument whose configuration and markings allowed one to replicate those angles and lengths, Sarpi had gestured vaguely to an optical exercise, suggesting that altimetry could take place "with the mirror," before turning to the issue of the camera obscura. Put differently, whereas Sarpi's notes preserved the experimental context from which some form of a telescopic instrument emerged, Galileo's sector, designed for a variety of uses, did not depend upon cast shadows or mirrored images of the objects under scrutiny and would have been entirely incommensurate with the project of enhanced vision.

But what is interesting about Capra's clumsy appropriation of Galileo's instrument is that he or his coauthor Simon Mayr evidently believed that there *were* hidden means of measuring, beyond the rules laid out in the treatise, and that these methods constituted a marvelous secret and a presumed improvement over the available device. Capra or Mayr or both were probably encouraged in these conjectures by Galileo's suggestion, in the closing pages of his treatise, that there were other ways of using his device to "measure with the eye," and by his bland promise to publish a fuller description of these procedures, and of the construction of the instrument itself, in the future.[40]

Because Mayr had left for Germany in 1605, the hapless Capra bore most of the burden of Galileo's indignation and blunt humor. His reaction to Capra's insinuations about his secrecy, his *Defense against the Slander and Pretensions of Baldessar Capra,* was a sustained effort to depict that young man as not merely too stupid to pull off a theft, but also tainted by his close association with someone "always engaged in reading diabolical treatises" and spewing toxins, a mentor who had repeatedly sought Galileo's downfall through other machinations or machines, and explicitly identified as Gromo in the last pages of the work. The several references to Gromo as an "ancient adversary," not just of Galileo but of all mankind, at once associate him with Satan, insist on a long-term enmity, and refer somewhat uncharitably to this rival's very advanced age.[41]

In some sense this last suggestion was overkill, as was Galileo's prediction that Gromo would bring a copy of Capra's confiscated work to Germany for reprinting, for the old man was in no shape for that or any other earthly journey. Even though Ingegneri had written in the course of his denunciation of his own past as an alchemist in August 1606 that the failed and penniless Gromo had died "just the other day" without producing gold or valuable medical remedies, and that he himself had accompanied him to the tomb, he also acknowledged that there were those who persisted in believing that the old magus would soon emerge from the crypt, rejuvenated, invigorated, rich, and vindicated. One can only assume that familiarity with this rumor accounts for two amusingly macabre touches in the *Defense,* Galileo's observation that virtuous people avoided someone charged with fraud and imposture as they would "not just unproductive trees, or even beggars, but also fetid cadavers," and his suggestion that in order to adjust the angle of his instrument in two botched measurements, Capra would need to heave himself into the grave.[42] As if to insist—though surely in jest—that Capra was still seeking some sort of direction from the late Gromo, Galileo elsewhere pointed to the young man's "diabolical conscience," suggested that he had used the sector to draw, or pretend to draw, not regular polygons but pentacles and other occult figures, and alleged that he had taken an unfamiliar term for an equilateral, *isopleuri latus,* for a frightening incantation.[43]

But it was in the matter of the secret means of measuring heights that Galileo took special care to distinguish his own intellectual habits from those of Capra and his elderly, otherworldly companion:

> If [Capra] wanted, as I believe, to indicate me as the person who keeps these means of measuring secret, he was truly mistaken, because if by "secret" he means something enormous and miraculous, as, for example, the secret of curing a wound from afar simply by anointing either the weapon that hurt someone or a bandage with his blood, or the secret of that astonishing ointment that makes even the stoutest iron blade crumple in a few hours, or other wonders of this sort, I not only have not regarded these rules of measurement as marvels, but I have always believed, and continue to believe, that such stunning things would not be found anywhere in mathematics.
>
> And if by "secret" [Capra] means something kept under wraps and hidden, he is even more mistaken, since I have never concealed these things from or denied them to anyone who asked me about them, which by now must be hundreds of gentlemen. And if finally by "secret" he means to suggest something new and unusual, I can certainly believe that many of my procedures are such, and above all those whose laborious calculations I myself have removed, such that they are resolved with just the compass and the Arithmetic Lines [that is, markings] I have provided, in a fashion never previously imagined by others.[44]

Galileo went on to say, not unreasonably, that the compass seemed arcane to Capra because of his ignorance of the most basic principles of the device, and it is clear that what really interested this young man was less the measurements provided by the instrument than what he imagined it allowed its user to see.

Capra therefore took the unnamed secret-holder to task for the scarcity of information about the *traguardo,* an accessory to align the sight; Galileo quite naturally objected that his enemy could not both claim the device as his own and fault others for failing to provide particulars about its usage, and he stated that all such details were addressed in his private lessons, without which the stolen manual was useless.[45]

To sum up Galileo's case against Capra, the numerous instances of the latter's ineptitude, as well as the affidavits from Paolo Sarpi, Gianfrancesco Sagredo, and Jacques Badovere attesting to the years Galileo had spent perfecting the device, easily convinced the authorities to seize all copies of Capra's work still within Venetian territory and to expel him from the university.[46] Though this outcome seems justified, the more crucial issues regard Capra's expectations, however unrealistic, of the device in its altimetric applications and Galileo's impression of the threat he posed. Given that Capra had claimed expertise in using an astrolabe in his observations of the New Star of 1604 in an earlier and slightly less daring treatise, it appears that his concerns with Galileo's "secret" had more to do with objects at a closer range. Because Galileo returned three times to Capra's irritating demand for more information about the traguardo, or sight, and referred no less than eight times to the fact that he did not align the instrument correctly with his eye, it seems likely that he understood that Capra's expectations involved an optical feature that this sector clearly never possessed.[47]

Despite the fact that Capra—a habitual offender—is among the most unreliable of witnesses, the suggestion that Galileo had arcane information useful to individuals engaged in measuring heights is an interesting one.[48] This peculiar inference was most likely determined by the expectations that had accompanied Abramo Colorni's unrealized rediscovery of instruments for measuring distant objects. It is no accident that many of Galileo's statements about Gromo's necromantic interests were the sort of thing said about Colorni, primarily because of his translation of the best-known occult text, the *Clavicula Salomonis,* for Duke Guglielmo or Vincenzo Gonzaga. If Gromo had spent all his time, as Galileo alleged, reading "diabolical treatises," it was Colorni who had made available an Italian translation of the most significant such work.

Though the dynamics of Galileo's case against the hapless Capra and the defunct Gromo meant that he could easily present their insinuations about his procedures as the corollary to occult dabbling and a kind of knavish stupidity, there is the strong possibility that the obsessive interest in the traguardo had to do with its combination with the concave mirror and their rumored application to telescopic vision. This preoccupation, more than Galileo's depiction of Capra and his "ancient adversary" as bumbling occultists, would explain the astronomer's occasional reference to the younger man's access to some aspect of his private lessons, which we know to have involved optics in 1601 and which Ingegneri's poem of 1602 suggests had a specific connection to the concave mirror. It is also clear, from his evolving position on the moon's secondary light and as a result of his investigation of different sorts of celestial light in 1606 and 1607, that Galileo was studying both reflection and the effect of restricted apertures on vision. It is likely that some of this research would have been available to Capra, whose initial contact with Galileo had to do with the New Star of 1604.[49]

Given that Gromo's "secrets" were in large part borrowed or adapted from della Porta's *Natural Magic,* it is reasonable to assume that he understood from book 17 of that work that telescopic vision in some way involved a combination of concave mirrors and lenses, and it is possible that he charged Capra with learning the particulars through his contact with Galileo and his private students. One of the more peculiar insults aimed at the late Gromo—that he was like the mythical basilisk—fits into the logic of an unarticulated quarrel about optical instruments with telescopic properties and mirrors as components. The basilisk, as Galileo acknowledged, was said to kill from afar with a glance; the corollary to that popular fable, which Galileo did not mention, was that it was in turn slain by seeing its own horrific reflection in a "great looking-glass."[50] Though by the time the *Defense* was written the dead man was a straw man, Galileo's anxiety about the improper disclosure of his ongoing work was genuine.

[Capra] emboldened himself to publish what he imagined that I said in my private lessons, and what I myself did not want to put in print. Now one will have to be very circumspect in the company of such persons, who, *like spies on the world,* very subtly gather together whatever someone else—either carried away in talking, or by accident, or even through ignorance—happens to say, and then *they convey it to the ears of the universe.* And will those privileges and abilities that posterity concedes to scholars so that they can become aware of their errors, correct them, review, polish, and revise their own writing one, two, or a hundred times, be thus abolished and annulled because of the petulant and vigilant censure of this man? I don't know in what schools Capra would have picked up such manners.[51]

The context is a preamble to Capra's impertinent criticism of Galileo's observations of the New Star of 1604, but the most telling features are the metaphors the astronomer has chosen to describe the plagiarist. These figures of speech recapitulate two functions traditionally assigned to the concave mirror in treatises such as Ausonio's *Theoretical Discourse,* that of telescopic vision, or espionage, and that of propagating sound. It is not necessarily the case that Capra actually possessed a concave mirror, but rather that he was aware of it as one component in ongoing experimentation with telescopic vision and needed to know more of the lens with which it would be paired.

The evolution of the word *traguardo* itself in this period also bears some traces of Capra's misguided assumption. Though it originally referred simply to a narrow opening where an observer or user of an instrument places his eye, the term briefly acquired the sense of a glass lens, and specifically the ocular or eyepiece, directly *after* the invention of the Dutch telescope. It was as if these speakers regarded the Italian verb *traguardare,* "to look through or across," as the vernacular equivalent of the Latin *perspicere,* which likewise meant "to look through" but was very strongly associated with enhanced vision and lenses, as the term *perspicillum,* an early name for the telescope, readily suggests.

Thus in the fall of 1610, after he had discarded his initial skepticism about the telescope, Giovanni Antonio Magini used the term *traguardo* interchangeably with "concave lens" in discussing the device with Galileo, and his associate Antonio Santini adopted this sense of the word in the same period.[52] In 1616, in one of his many attempts to ascribe a telescopic instrument if not to a Florentine at least to someone strongly associated with that city, Raffaello Gualterotti stated that there had been *traguardi* in the form of lenses in the late sixteenth century. He described an instrument designed by Egnazio Danti for Cosimo I de' Medici and thus made before the Grand Duke's death in 1574, saying that it was an enormous folding rule of brass, mounted very high on a wall and used daily to measure polar altitude with great precision, and that it was equipped with large lenses in order to make such minute measurements visible to onlookers below. No evidence of such an instrument exists, though it is true that the armillary sphere and, to a lesser degree, the quadrant Danti had placed in another location in Florence were too high above the ground to be easily viewed.[53] Gualterotti's additional assertion that Galileo himself wanted to update this phantom brass rule may have more to do with his own imaginative tendency to retrofit older instruments in order to ensure the Florentine provenance of the Dutch telescope, and it is surely significant that the device he described is an oversize version of Galileo's geometrical compass.

There is no reason to suppose, however, that these connections could have been drawn only by those familiar with the rumors about Colorni's instruments, for as Chapter 2 showed, many of the same suggestions had been made a quarter century earlier in the English context in Edward Worsop's dialogue on surveying. There, in the wake of the Baconian device presented in Digges's *Pantometria,* a Catholic priest with what both seems and seems not to be occult gear carries out surveying practices, looks through an ambiguously described instrument to see something rather remote, and writes down a few formulae of unknown content. The misguided assumption that a surveying device might offer enhanced vision—conveniently voiced by a naïve speaker whose unvarnished observations are never fully endorsed or discarded in the course of the dialogue—is precisely the same "secret" sought out by Capra and his associates.

In sum, though it is clear that Galileo's compass had nothing to do with an optical device, it does seem likely that Capra's extraordinary interest in the traguardo had to do with both a general awareness that discussion of a telescopic device had emerged within the context of altimetry, and his specific impression that Galileo was investigating some combination of a concave mirror and lens. Capra's questions about further refinements to the compass, the focus of such rhetorical energy in the *Defense,* had a certain afterlife. The whole episode was presented by Galileo's earliest biographers as a prelude to his subsequent rise to international prominence through his deployment of the Dutch telescope and extraordinary celestial discoveries. These accounts alter the gradient of his ascension in that they suggest that his geometrical and military compass was already well known throughout Europe, and they soften the familiar image of the astronomer as the bold appropriator of others' devices by showing him first as the victim of an audacious thief.

Vincenzio Viviani's biography is overall the more informative and reliable one; however, the less-polished contemporaneous version by Niccolò Gherardini offers a far better approximation of the popular understanding of telescopic vision in this period, of Galileo's alleged role in the rediscovery of an ancient instrument, and of the astronomer's own anxiety about Capra's disclosure of some imperfect or incomplete features of his research. Stating that Galileo welcomed detractors like Capra because their attacks gave him the opportunity to improve weak or unclear points in his work—an observation indignantly contested in one of Viviani's marginal notes—Gherardini went on to praise the philosopher's decision "to renew for the world the means, long abandoned and no longer even hoped for, by which an instrument might be made that could so enhance one's visual faculty."[54] Written around 1655, Gherardini's narrative was already a peculiar version of events, but it perfectly catches the mood, and something of Galileo's actual research, in 1607.

Chapter Four

The Dutch Telescope and the French Mirror

SOMETIME in November 1608, Fra Paolo Sarpi encountered *The Embassy of the King of Siam Sent to His Excellency Maurice of Nassau,* a French-language news pamphlet from The Hague describing the first visit of the Siamese to Europe, portraying the gifts their ruler had sent abroad, offering a brief sketch of the wealth, culture, religion, and political structure of that remote kingdom, and alluding to the commercial inroads made by the Dutch East India Company, largely at the expense of the Portuguese, in the Far East.[1] The news of the expedition, with somewhat different accounts of the gifts involved, had already emerged in manuscript newsletters sent from Antwerp on September 19, but the addendum to this pamphlet mentioned a slightly more recent event, the invention of the refracting telescope in the Netherlands.[2]

Sarpi's extant correspondence suggests a wary skepticism about the instrument, and in the succeeding months he was quick to inform others that the Dutch device was already old news to him and even quicker to point out that, as a former student of optics, he was unwilling to discuss the invention before trying it out. As dismissive as such remarks appear to be, they are also accompanied by a certain nostalgia. Sarpi alluded with studied vagueness to his early experiments with an optical device of parabolic shape, and he wrote to the Frenchman Jérôme Groslot de l'Isle, "I don't know if that artisan [from Middelburg] had my idea, or whether the whole thing did not acquire magnification, as rumor always does, over the course of its journey."[3]

In early December Sarpi wrote with the same measure of languorous caution to the Huguenot Francesco Castrino that he had received the report from The Hague a month ago, "but because these philosophers bid us not to reflect upon the cause before witnessing the effect with our own senses, I have gone back to waiting for such a noble matter to spread throughout Europe."[4] Though he made no explicit allusion to his own investigations of telescopic devices, his choice of the verb *specularsi,* "to reflect upon," like the reference to the parabolic instrument, suggests that Sarpi assumed that the Dutch device involved a mirror.

In this same period he reacted with equal skepticism to another recent publication, Johannes Cambilhom's *Discoverie of the Most Secret and Subtile Practises of the Iesuites,* mentioning it dismissively to Groslot in his letters of September and November 1608. Originally composed in Latin and rapidly translated to German, Italian, French, Dutch, and English, this pamphlet described the many misdeeds of the Society of Jesus, most notably its bawdy adventures with innocent girls, its vast stores of buried treasure, its arsenals of weaponry, its sadistic treatment of recalcitrant novices, and its grandiose political designs. An ex-Jesuit himself, the Austrian Cambilhom promised but never delivered a fuller treatment of his subject in the future.[5] Sarpi assured Groslot in mid-September that he had no doubts about the nature of the Jesuits, but added in regard to their "arcane doings," "We've certainly not had a whiff of such things here in Italy." Returning to the issue some two months later he complained, "Being most subtle masters in evil, it is credible that the [Jesuits'] arts are as various as the regions in which they operate. But if the author of this booklet will confirm what he says with examples that make the truth manifest, it will be of universal benefit."[6]

What is particularly interesting about Cambilhom's *Discoverie,* which otherwise is confined to an examination of Jesuit practices in Austria and Bohemia, is its revelation about an optical device owned by the French King Henri IV's confessor, Father Pierre Coton. "The Iesuits them-selves brag that hee hath a looking glasse of Astrology [*speculum constellatum*], wherein he made the King to see playnly what-soever his Maiestie desirded to know, and that there is nothing so secret, nor any thing propounded in the privy councells of other Monarkes, which may not be seene or discovered by the meanes of this celestiall or rather divilish glasse."[7]

What was the exact connection between the Dutch spyglass and the French mirror? News about the latter appears to have played a role in Sarpi's and Galileo's first and erroneous impression of the refracting telescope. In particular, both of these instruments—that familiar icon of the Scientific Revolution, and this bogus bit of Jesuit devilry—must be understood within a traditional context, one where tales about telescopic devices were both a common literary motif and to some extent a correlate of recent catoptrical experimentation.

The conventional account of Galileo's belated awareness of the telescope—that he heard nothing of it for eight or nine months after its emergence in The Hague—must therefore be reexamined in light of the news concerning the mirror in Paris. Both Galileo and Sarpi seem to have known of the existence of the Dutch invention by November 1608, and it appears that their familiarity with catoptrics predisposed them to imagine a device that involved a lens-and-mirror combination, rather than a weak convex lens combined with a strong concave one within a tube. The role of Jacques Badovere, the man named in *Starry Messenger* as Galileo's eventual informant, also merits new attention. Galileo's and Sarpi's efforts to obtain relevant details about the telescope from their friend in Paris were determined not just by his expertise in optics and his circulation in diplomatic circles, but also by his status as Father Coton's protégé and confidant.

Jesuit Prescience

Well before *A Discoverie of the Most Secret and Subtile Practises of the Iesuites* disclosed the news of Father Coton's *speculum constellatum* or "starry mirror," members of the Order were rumored to have exceptional means of knowing the business of others. In 1594, for example, in the *News from the Regions of the Moon,* an anonymous and very popular French work associated with Sarpi's friends in Paris, the Jesuits were satirically portrayed as colonizers of the moon, and as creatures "almost all of whom have fox tails attached to their belts, along with mirrors, with which they see what is done in the world, and dazzle the eyes of those who look at them." The concave mirrors that the Order allegedly used to scrutinize others and to blind their enemies turn up elsewhere in the *News,* for the narrator also alludes to an elite "who use burning mirrors, igniting all sorts of wool through reflection of the sun's rays."[8]

Even though such a portrait of the Jesuits is clearly apocryphal, its insistence on the Society's study of mirrors is not misplaced. It was no accident that when Gabriel Naudé sought in 1625 to free Roger Bacon of the charge of magic, he chose to defend the English friar's legendary skill in catoptrics by pointing, with heavy-handed irony, to the Jesuits' recent exploits in the same discipline:

> For being such a great mathematician, as one can see, as much by the treatises and the instruments of [Bacon's] own invention sent to Pope Clement IV as by those two books of his, published just a decade ago, on perspective and mirrors, we can well believe that he managed to do many extraordinary things with this knowledge. The underlying causes being unknown to the common people—who were then much more primitive and barbarous than those of today—they could not help being associated with magic. And yet I believe he will always be supported by learned men, and above all by the Most Reverend Fathers of the Society of Jesus, who have not neglected to mention in the Theses in Mathematics that were defended in Pont-à-Mousson in 1622 on the day of the Canonization of Saints Ignatius and Xavier, that it was possible

for a man well-schooled in optics and catoptrics—as Roger Bacon undoubtedly was—"if given any object whatsoever, to show anything at all in the mirror, as for example a mountain from an atom, a swine's or ass's head from a human's, or an elephant from a hair."[9]

Jesuit studies of burning mirrors range from the sober work of Christoph Grienberger and his student Francisco Guevara in 1613 to the more fulsome *Apiaria* of Mario Bettini of 1642, and finally to the spectacular war game featured in Athanasius Kircher's *Ars Magna Lucis et Umbra* of 1646, and in the triumphal iconography of Sant'Ignazio in Rome, where an angel holding such a device ignites its terrestrial target.[10]

The mirror-bearing Father Pierre Coton, the subject of Cambilhom's news pamphlet, had become the focus of rumors devoted to Jesuit prescience around 1605, and these charges appear to be the special concern, or rather invention, of Paolo Sarpi's French contacts. Coton was said, for instance, to have sought out information about different European monarchs from an unclean spirit encountered in an exorcism he performed, and the brazen questions he allegedly asked soon found their way into manuscript and print.[11] According to the French chronicler Pierre de l'Estoile, Father Coton's *grimoire,* or manual of black magic, enjoyed a scandalously wide circulation among the populace in the fall of 1605.[12] In 1607 Coton, said by his enemies already to "govern both heaven and earth," was mockingly implored in the *Passport of the Jesuits* to "content yourself with that . . . For instead of making a devil talk, you have set the whole world gossiping."[13] And as late as 1624, letters by Sarpi's friend Jacques Gillot describing how Coton's diabolical queries had fallen, "as if by miracle," into his hands, were published in Paris.[14]

In 1606 Father Coton was harassed by an astrologer of some renown, a man who brazenly professed himself his brother-in-law and who sent him letters emphasizing this baseless parentage. The astrologer's assertion was publicly denied by Coton, who insisted that this stranger both abandon every such claim and renounce all future activity in judicial astrology, horoscopes, and physiognomy.[15] Within eighteen months, however, Cambilhom's *Discoverie* appeared, its revelations about Coton's starry mirror clearly stimulated by previous rumors associating Jesuit acumen with devilry. As Cambilhom explained, the Society of Jesus as a whole was given over to the study of black magic, and in this area Coton "excells all those of his sect. The French king did so much esteeme him as hee did assist alwaies at his Table, and did commonly intertayne him: wee have seene the questions which hee propounded in the yeare one thousand six hundred and five, to a young Mayden that was possest at Paris, whereby his wickednesse, and the pleasure hee takes to speake with the Divell appeareth playnly."[16] By late 1610 Coton was identified as a kind of glib magus whose special interest was astrology or astronomy; he was said to sprinkle his conversation with "a few choice words taken from [Sacrobosco's] *Sphere,* like 'poles,' 'tropics,' 'apogee,' 'ascendant,' 'zenith,' and others like it, [terms] with which he mocks and makes himself admired by foolish courtiers."[17]

Of the many rumors circulating about Coton, it was the business about the mirror that particularly intrigued the numerous enemies of the man and his Order. When the chronicler Pierre de l'Estoile was loaned a copy of Cambilhom's *Discoverie* in January 1609, he had time only to copy out the passage about this instrument and to note that some believed the work to be accurate, for all the apparent clumsiness of its execution.[18] The anonymously published pamphlet was almost immediately translated to French, and both versions were passed around in clandestine fashion, no one wanting to take responsibility for producing, owning, or even reading anything so hostile to the king's confessor and to his fellow Jesuits.[19] Paolo Sarpi's copy, in fact, had been conveyed to him in a diplomatic pouch sent by Antonio Foscarini, who routinely used his position as Venetian ambassador to France to circumvent censorship and unreliable postal services, and the theologian promised to show it only to trustworthy persons.[20]

Not surprisingly, the bookseller who had printed the French translation of the work was jailed in September 1609; as l'Estoile then saw it, the *Discoverie* was "pure nonsense" not worth the paper on which it was written, but the implication of Henri IV in the story about Coton's mirror made exemplary punishment for the imprisoned man inevitable.[21] The censorship of the *Discoverie* revived the story of the diabolical questions: in mid-October 1609 l'Estoile was also presented with what he was told was a most accurate copy of the queries, "much more exact than those in circulation," detailing the Jesuit's research into "divine, natural, political, celestial, terrestrial, infernal, and sluttish affairs," and insisting on the transmission of the incriminating document from Jacques Gillot to Henri IV's most powerful minister, the Duke de Sully.[22]

Most of the foregoing rumors would have been familiar, if not very credible, to Paolo Sarpi by late 1608, when he sought to inform himself about that other optical device, the telescope lately invented in the Netherlands. He surely knew the absurd detail in the *News from the Regions of the Moon* about the infernal beings with fox tails and concave mirrors, for from 1594 this text was always printed as the continuation of the *Menippean Satire of the Spanish Catholicon,* a work reputedly written in the home of his close friend Jacques Gillot and often mentioned by Sarpi.[23] Gillot, of course, was not coincidentally the man who allegedly found the list of questions posed by Pierre Coton to the exorcised demon, and the chief purveyors of this tale, the Duke de Sully, the historian Jacques-Auguste de Thou, and the chronicler Pierre de l'Estoile, were also in sympathy and close contact with Sarpi. Whether or not the Venetian friar had heard the bizarre story of the astrologer who claimed to be Coton's brother-in-law, he would have readily understood the pretender's intention as a desire to enter into the Jesuit's presumed orbit of foresight and far-reaching knowledge.

Sarpi certainly knew that there was little truth in these matters, for he not only collected anti-Jesuit pamphlets but is thought to have been the anonymous author of a few such publications in these years.[24] That said, it is likely that he regarded the report about the optical instrument offered by the "humble and God-fearing" lens maker to the Dutch ruler Maurice of Nassau in order that he might see clock towers and church windows a few miles away as a tame version of the contemporaneous account of the talismanic mirror with which Father Coton showed the

French king all that took place "in the privy councells of other Monarkes." Put differently, Sarpi associated the two stories, considering them both exaggerated accounts of the special effects that might be achieved with mirrors and of the utility that such instruments would have for rulers. It is no accident that he and Galileo regarded Father Coton's close associate, Jacques Badovere, as a likely source of information.

The French court, finally, was not indifferent to recent developments in catoptrics in Italy, and Queen Marie de Médicis, having left Florence to marry Henri IV only in 1600, was strongly associated with such interests. She commissioned an Italian translation of Ettore Ausonio's *Theoretical Discourse on the Concave Spherical Mirror* sometime between 1602 and 1609, and she, like many other early modern rulers, was often approached by those who sought to sell her large and small looking-glasses.[25] That mirrors were prized both as costly playthings and as objects of speculation was almost certainly the case at the court. The first preceptor of the Dauphin Louis XIII, Nicolas Vauquelin des Yveteaux, dismissed for a certain frivolity in pedagogical matters in early 1611 after less than two years' employment, would later explain the inclusion of romances and poetry in the royal curriculum with a telling analogy: "So that one comes to welcome the sight of magnificent temples and soaring palaces, [visits to] grottoes must not be forbidden, and neither mirrors nor perspective glasses treated with contempt."[26] And David Rivault, an important presence at the court since 1603 and the Dauphin's preceptor in mathematics from 1611 onward, would offer in his translation of Archimedes' work in 1615 by far the greatest impetus to the legend of the events at Syracuse by maintaining that the context proved that burning mirrors, rather than fiery pitch or other less glamorous flammable materials, had been involved.[27]

The Most Illustrious Signore Jacques Badovere

Jacques Badovere was the French-born son of a rich Venetian émigré whose fortune had been destroyed, like that of so many Protestants, in 1572 during the Saint Bartholomew's Day massacre in Paris.[28] His friendship with Sarpi and Galileo dates to his studies in Padua from 1597 to 1599, when he lived in the astronomer's house, enjoyed the company of the Venetian friar, and participated, like Gianfrancesco Sagredo, in studies of the geometrical compass. In these years he was also in close contact with the Dutch humanist and poet Pieter Corneliszoon Hooft, who had settled briefly in the glassmaking center of Murano prior to roaming about Italy, and who remained a correspondent for at least the next decade.[29] Badovere traveled with Hooft from Florence to Venice in mid-October 1600 and undertook transactions regarding his late father's estate in Venice in January 1601; in this period he also investigated magnetism and tidal theories, subjects pursued by Galileo, Sarpi, and Sagredo since the late 1590s.

Badovere returned to the French capital around September 1601 and by 1602 or 1603 was serving King Henri IV as a secretary, being entrusted with the delivery of his letters in Germany and Austria.[30] Under the influence of Father Coton, who was then engaged in reestablishing the Jesuits in France, Badovere embraced Catholicism; he described his return "to the former and fortunate belief of my forefathers" as a recent event in a letter of February 1604, though documents of 1610–1612 suggest that he might have collected several *pensions* in those years for abjuring Protestantism.[31] As late as 1617 his conversion still troubled Sarpi, who attributed Badovere's decision to the Jesuit's undue influence.

> I knew Jacques Badovere for a long time in Padua and Venice, and he was devoted to the Reformed Religion to the point of superstition. Yet when he went back to France, he fell away to our faith. When he came back to Italy [in 1607], I asked him for what reason he could have abandoned the faith of his ancestors, the one in which he was born and brought up, and he answered that Father Coton, having passed through Melun or rather Abdera [a Thracian city noted for its foolish inhabitants], used compelling arguments to uproot and erase all religion, and then implanted in one's empty breast the most useful one. What might I not fear for you from this man who fears no God?[32]

Badovere remained in contact with and occasionally visited Sarpi and Galileo, and in the summer of 1607, when he sought to establish his priority in the invention of the geometrical compass, Galileo called upon Sagredo, Badovere, and Sarpi for attestations in his favor.[33] Although this choice had to do with their familiarity with that device, their strong social standing, and their presence in the Veneto, it is feasible that Badovere, like Sarpi and Sagredo, had some notion of Galileo's study of optics. Though in January 1610 Badovere would be described in scurrilous verses as the puppet and spy of Father Coton—the latter allegation finding some support, perhaps, in the fact that he forwarded at least one of Sarpi's letters to the Jesuit priest and to the papal nuncio Robert Ubaldini around July 1609—Galileo and his former student retained a lifelong devotion to each other.[34]

That Badovere, like his late father, was tolerably well connected in diplomatic circles can be only one reason, and perhaps not the most important one, that Galileo and Sarpi turned to him in search of information about the news from The Hague. Sarpi mentioned Badovere several times in letters to Antonio Foscarini, the Venetian ambassador to the French court, but each allusion to his talents was accompanied by a word of caution.[35] In August 1609, when Badovere was entrusted by Henri IV with diplomatic functions in a succession crisis in the Rhenish territories of Clèves and Julich, Sarpi foresaw the success of his "impatient nature" only if proceedings were rapid, and in fact the French minister Villeroy and the Duke de Sully, alarmed by Badovere's close connections to Father Coton, promptly ensured his recall and professional humiliation.[36] And when Badovere was described in a pro-Jesuit pamphlet of July 1610 as "as knowledgeable in foreign as in domestic affairs," Sarpi concluded that this was no evidence of the Frenchman's influence on European politics, but rather proof that Badovere was the actual author of the text.[37]

It was thus perhaps as much Badovere's close friendship with Father Coton as his access to those involved in the negotiations in The Hague that interested his Italian correspondents in late 1608. What is known of Badovere's other contacts and interests also makes him a credible messenger for technological developments. In a letter of December 1607, just after returning from Padua, he begged his friend Hooft, then living in Amsterdam, to provide him with news right away and to let him know the fate of certain books Hooft had sent from the Netherlands "about statecraft, commerce, navigation, warfare, hydrology, machines, and other useful and curious things." He also asked Hooft to tell him whatever he knew about "military secrets, including the one that allows men not enclosed in an apparatus or a ship to move about and to stay under water," and in a letter of mid-February 1609 he referred to his interest in "perpetual motion machines and similar things."[38] Badovere's letters in this period also suggest that he was in close enough contact with two men associated with the Dutch telescope—Pierre Jeannin and François d'Aerssen— that he named them as trustworthy recipients of a large sum of money owed him.[39] And in the spring of 1612 he visited the poet and glassmaker Girolamo Magagnati in Venice and, according to Sarpi, was in constant conversation with alchemists, though perhaps as a cover for other and worse activities.[40]

Spyglasses and Letters

The fragmentary letter collections of Sarpi and of Galileo offer some of the best evidence for their first impressions of the telescope. Sarpi's activities and his correspondence were in these years the subject of enormous scrutiny in Rome, where Scipione Cardinal Borghese, nephew of Paul V and papal secretary of state, managed to obtain copies of many of the letters Sarpi sent to France and to have their contents reviewed by both the pope and Henri IV. Cardinal Borghese coordinated his efforts with the papal nuncio in France, Roberto Ubaldini, who enlisted a trio of helpers in this project. Ubaldini convinced the secretary to Antonio Foscarini, the Venetian ambassador to France, to obtain Sarpi's letters to the latter, and he had a Bolognese resident in Paris copy Sarpi's letters to the Huguenot Francesco Castrino, and he also persuaded Jacques Badovere to divulge the contents of at least one of the letters he had received from the theologian, that involving the telescope.[41] Sarpi had been warned by others that his correspondence was monitored, and it is possible that these suspicions account for the somewhat uncommunicative and occasionally misleading tenor of his letters in this period.

Although it would be wrong to assume that the optical information in his correspondence was what most interested Sarpi's enemies, the theologian himself seems, perhaps inadvertently, to have created that impression. His friend Jérôme Groslot de l'Isle, who had first mentioned both the Dutch telescope and the story about Father Coton's telescopic mirror in the fall of 1608, enclosed a separate report about the telescope early in 1609, and referred to the deployment of the device as a "miraculous event" in his own letter. Having opened several missives all at once, Sarpi assumed that this account came from Antonio Foscarini, the Venetian ambassador to France, and he replied to Groslot that he had failed to understand his references to the miraculous enclosure. In his response, Groslot conjectured that the news item had been maliciously removed from the letter, a hypothesis Sarpi finally laid to rest in mid-May in a brief explanation of his error, accompanied by the usual profession of skepticism about the device. What is most interesting about the whole event is the fact that even in mid-May Sarpi appeared to connect the telescope with the news about Father Coton's mirror, for after writing that he could not comment upon an instrument he had not yet seen, he added, "But when Your Lordship takes me from this miracle to that other wonder of the Jesuits, I can certainly say that it is something that I myself have seen and known, though not fully. They have so many hiding places, so many pretexts, and so many disguises."[42]

Sarpi's letter to Jacques Badovere in late March 1609 is, by comparison to his exchanges with Groslot, even less straightforward, though some of its apparent obliquity may be due to the fact that it is a reply to an inextant letter from Badovere where the telescope had already been mentioned. Sarpi began with an admonition about Badovere's health—which had evidently been ruinous during his trip to Italy in 1607—stating that effective cures take time and patience, and implying that he knew the necessary repose and calm would not be to his addressee's taste.[43] The friar then proceeded to the matter of the telescope itself:

> As for the Dutch glasses, I have already told Your Lordship my opinion, though it could be entirely wrong. If you were to learn anything else, I would gladly hear what people think where you are . . . Your Lordship tells me of the workings of the sense [of sight] only, a thing incomprehensible to me, since I do not see how [just] one part functions. Until now I believed that every operation involved the entire being, just as it does not appear to me that my hand alone writes, but that I apply myself entirely to the task, even the legs. These are my excuses for distinguishing between sense and theory.[44]

This mild complaint suggests that Badovere's description of the telescope had emphasized only the sensory impression that the instrument made—the fact that it magnified objects five or so times, for instance—and relayed little or nothing about the physical or theoretical means through which such effects were achieved. Such information would have been crucial given that Sarpi still believed in late March, as he did in early January, that the telescopic device involved a lens-and-mirror combination. Badovere might also have implied in his letter that he believed that Sarpi was again conducting optical experiments with Galileo, but his addressee set that suggestion aside in his reply by stating, "As for what Your Lordship desires me to tell you, I suspect that you believe me much more a participant in this business that I am." He assured Badovere that his "chimerical conclusions" would only make him laugh.

A particular detail in Sarpi's letter merits careful consideration. Though the theologian implied throughout that he no longer pursued his studies of optics, the suggestion itself, made in the midst of his statements about how little he knew of the telescope, is entirely undercut by the resonance it gains from a famous classical precedent. "I have almost abandoned the study of natural and mathematical matters," Sarpi insisted, "and to tell the truth, my mind has become, either through old age, or through habit, a little obtuse for such thoughts. Your Lordship would not believe how much I have lost—in bodily health as well as in spiritual well-being, and in mental acuity—as a result of this political song and dance."[45]

The statement, which Sarpi also paraphrased in his *Pensieri,* derives from Seneca's *Letters to Lucilius,* and it concerns the sorts of lies one should tell in order to disguise, rather than to display, one's philosophical activities: poor health, mental incapacity, and a professed tendency to idleness were all useful excuses when one had study to pursue. As Seneca put it, "You should not attribute the cause of your retirement to a desire to attend to philosophy and contemplation. Call your plan by a different name: say it is physical and mental frailty, and worse still, inactivity."[46] Badovere, whose letters make evident his strong classical education, would have recognized the fiction for what it was, and would have seen it as an unequivocal affirmation that Sarpi *was,* in this period, pursuing natural philosophy and investigating the optical tool he appeared to neglect.

One of the most interesting passages of Sarpi's letter runs as follows: "Concerning the business about the wells, what I have always believed, and continue to believe, is that it won't work, but reason must give way to experience. When I see it done, I'll say that it is doable, but not before."[47] This reference probably involves Sarpi's alleged distrust of the Aristotelian notion that an individual standing in a dry well would be able to observe stars during the daytime.[48] As Chapter 3 showed, in the 1580s Sarpi certainly did believe in the closely associated phenomenon of the cerbottana, where a tube without lenses produces slight magnification of nearby objects. Though it is surprising to find no allusions to the more easily verified issue of the cerbottana in the company of "the business about the wells," in fact Galileo's recent research had compromised an important aspect of that argument. Galileo had found that in the case of small bright objects such as stars, their irradiation made them appear larger and thus closer to the naked eye than they actually were. A restricted aperture, like that of the cerbottana or the camera obscura or the eventual telescope, would strip away an illusory crown of rays and show the object as diminished in size.[49] There would be no telescopic effect, and the neat continuity between the tube and the impractical well would be undone.

More telling than Sarpi's reaction, which consists of his habitual profession of reluctance to believe what he has not seen, is the fact that Badovere had apparently mentioned the viewing well in his letter about the telescope. Such a reference perhaps indicates that he knew no more than that the Dutch telescope involved a tube, for this detail would naturally have been offered by anyone who saw the instrument and was not intent on obscuring its design. In any case, Sarpi followed "the business about the wells" by informing Badovere that he had sent a letter to someone named "Valamens," and that the response would be included with his own. Valamens, or Van Lemens, was Badovere's courier; the Frenchman would use him to contact Galileo about a year later in an effort to persuade the astronomer to find an extraordinary star—something equal to Jupiter's satellites—for the king of France, though by the time the letter arrived, Henri IV had been assassinated and Galileo had no need to attend to this rather awkward request.[50] Sarpi's statement strongly suggests, therefore, that he relied on Van Lemens to deliver a letter from Badovere to someone nearby, but not necessarily in Venice, and that he believed that a response would be forthcoming very soon and presumably available to him. As subsequent correspondence shows, Badovere's other addressee was Galileo.

This letter, like so much of Sarpi's correspondence in this period, was scrutinized by the papal nuncio in France and forwarded to both Henri IV and Pope Paul V. The French king, for his part, was not much engaged by what Sarpi, or any one else, had to say of the telescope; the one he had been given in January 1609 had barely caught his attention. The pope's nephew, Cardinal Borghese, would receive a telescope from Cardinal Guido Bentivoglio, one of Galileo's former students, early in the summer of 1609, but it is certain that his interest in Sarpi's letters had much more to do with religious and political differences. The papal nuncio in France, Roberto Ubaldini, in sending Cardinal Borghese a précis of this letter in July 1609, appears not to have recognized the allusion to Seneca's letter about disguising investigations in natural philosophy, concluding instead that the theologian was "sorry to have become involved in this [political] mess, and that not a day passed by that he didn't long for his former leisure, and that if he weren't so old, he would go to France."[51]

Those Enemies of Ours

Though news of the Dutch telescope reached Sarpi in November 1608, the first mention of it in Galileo's letters occurs in late August 1609, in his announcement that he had devised such an instrument himself. His rather sparse correspondence for the intervening nine or so months concerns the precarious finances of his brother-in-law in Florence, a horoscope he was asked to prepare for the Grand Duchess Cristina of Tuscany, the death of her consort Ferdinando de' Medici, the parabolic path of projectiles, his great desire for more free time for investigations in natural philosophy, and an irritating change in his teaching schedule in

Padua.[52] Though these letters lack, for the most part, the energy and specificity of those written in the course of the previous year about magnetism, in two missives of February 1609 Galileo alluded with more enthusiasm than detail to "three or four conclusions or effects that I have observed and already demonstrated, ones that will perhaps surpass in marvel the greatest curiosities so far undertaken by men," and to several findings, one alone of which, he stated, would guarantee his financial future, if it caught the notice of "a great prince who took pleasure in it."[53] It is likely that one of these projects had to do with mechanics, but the other endeavors remain unidentified.[54] Given, however, the repeated allusions to his interest in concave mirrors in 1601–1607, his fear of some disclosure about his project by Capra, and the clear indications that Sarpi associated the Dutch device with a telescopic mirror from the fall of 1608, it is not unreasonable to assume that one area of study was catoptrics.

Galileo's conviction that the publication of his work on these unidentified projects would bring high praise for both him and a patron, and "something of greater, broader, and more constant utility to scholars than anything [he] might do in the rest of [his] life," while underestimating his future accomplishments, is remarkable in that it anticipates, by a little over a year, the impact of his *Starry Messenger*. Comments made in a letter written by Gianfrancesco Sagredo, who had left Venice in the spring of 1608 to take up a position as Venetian consul in Syria, help establish some of Galileo's activities in this poorly documented period by showing, first of all, that he had attempted to contact Badovere late in 1608.

Shortly before his departure, Sagredo had carried out an elaborate practical joke on the Jesuits—an epistolary hoax where he pretended to be a rich, pious, and foolish widow in need of a confessor from that Order—and he was in the process of organizing another prank to take in Jesuit missionaries as well; though he alluded to both schemes in letters to Galileo, he found in Sarpi a more appreciative auditor.[55] Sagredo's reply to an inextant letter written by Galileo on April 4, 1609, or less than a week after Sarpi's letter to Badovere, is marked by his particular antipathy for the Society of Jesus and his general hostility to extraordinary displays of devotion, but it is most noteworthy in what it conveys about Galileo's activities in late 1608 and early 1609.

> In replying I will skip the ceremonious greetings Your Excellency used with me in your letter of April 4, which reached me by way of Constantinople on September 16, both because I am in a hurry, and also to show you that as regards that other business, you should not waste time in these unnecessary things.
>
> The success that Your Lordship did not have in communicating with your student, now that it has been conveyed to him by word of mouth, will have perhaps given him a sufficient taste and warning enough that he will know and guard himself against those enemies of ours. Their habit of making every day a holy day, what with their vespers and final evening prayers, has a certain resemblance to the superstitious practices of those here in this country, who repeat their hymns five times a day.[56]

This passage concerns Badovere, whom Sagredo knew well and whose ultra-Catholic piety he, like Sarpi, found repellent. Sagredo's comparison of Jesuits and Moslems, moreover, has something of an analogue in Sarpi's correspondence: in a letter of 1610, for instance, the theologian would note in the margin, "Jesuits are similar to Janissaries," the elite military and political group on which the Ottoman Empire depended.[57]

Most importantly, Sagredo's remarks supplement the basic timeline established by Sarpi's correspondence: Galileo seems to have made an attempt to reach Badovere in a letter that miscarried, though not necessarily through any malfeasance on the part of the Jesuits as Sagredo implied. Such efforts would have to have been in late 1608 and presumably would have had something to do with Galileo's two feverish references to his promising new projects in February 1609, and perhaps to his otherwise rather banal description of the late Grand Duke's prudent governance as "a mirror for other potentates" in this month as well.[58] Sarpi, using the diplomatic courier, let someone in Paris—presumably Foscarini, who saw Badovere frequently—know that Galileo was seeking contact with his former student. The latter's responses reached both men, with the aid of the courier Van Lemens, in mid- to late March 1609. These letters did not contain precise information about the Dutch telescope, and both Sarpi and Galileo, to judge by Sagredo's response, were frustrated by this delay. As agreed, Van Lemens carried their answers to Badovere, probably in early April, and the Frenchman evidently sent crucial details about the device back to Venice shortly thereafter.

That information distinguishing the Dutch telescope from the mirror allegedly deployed by Father Coton did not reach Venice until early in the summer of 1609 is strongly suggested both by the four- or five-week period most letters took in either direction and by the fact that Sarpi still associated the device with other Jesuit arcana as late as mid-May. The letter that reached Galileo in the summer of 1609 is the one whose author he acknowledged as "the noble Frenchman Jacques Badovere" in *Starry Messenger*.[59]

Ghost Images

Events moved quickly from the summer of 1609, for once he obtained the barest sketch of the Dutch telescope Galileo was able to make his own in short order. Even this initial prototype must have been superior, if only from the viewpoint of cost and utility, to anything he and Sarpi had contrived by combining a lens and a mirror. Though one or possibly two telescopes appeared in the Veneto by August, Sarpi managed to protect his friend's interests, and Galileo approached the Doge of Venice late in that month with news of a device "drawn from the most recondite speculations in optics," of great use in military and other situations, and "one of the fruits of the science which he had professed for more than seventeen years at the University of Padua." He suggested that he would be able to offer still greater things to the Senate, if so desired, and he soon saw a guarantee of lifetime employment at Padua and a handsome increase in his salary, albeit one that was both deferred for a year and presented as the last of such raises available to him.

Over the course of the fall of 1609, Galileo was able to improve the telescope significantly. The device he offered in late August to the Doge of Venice magnified eight times, but the one he had perfected three months later magnified twenty times, and with it

Galileo undertook the extraordinary series of astronomical observations that first brought him fame throughout Europe.[60] His *Starry Messenger,* emerging from the press of Tomaso Baglioni in Venice in mid-March 1610, began with a detailed comparison of the lunar and terrestrial globes, included a hasty and somewhat confusing sketch of a few constellations, and concluded with what was for natural philosophers, if not for the public, the most startling revelation, that of the "Medici stars," or the moons of Jupiter.

The timeline of Galileo's celestial observations and discoveries is now very well established. His lunar observations were made, for the most part, between November 30 and December 17, 1609; his attention to the startling pattern of celestial objects around Jupiter dates to January 7, 1610; and his decision to rush his work into print was made very soon thereafter. By January 30 he was in Venice conferring with Baglioni, who had printed his *Defense,* and he gave him the first part of the manuscript, which dealt with the rumor of the telescope and the lunar discoveries. Several weeks later he delivered some of the satellite observations, apparently arranging for Baglioni to leave a few folio leaves in which the constellations would be discussed and portrayed. He received the license to print *Starry Messenger* on March 1, made his last observations of Jupiter's moons on March 2, turned these over to the printer, and received an unbound and still damp copy of the work from the obliging Baglioni on March 13.[61]

The hybrid character of *Starry Messenger* is thus a function of both the different phases of its composition and the urgency that emerged with the discovery of Jupiter's satellites. The engravings of the moon depict what Galileo saw in late November, when he first took up his 20× telescope; however, the leisurely pace and philosophical tenor of the accompanying text, and the fact that many of the conclusions explained there occur in his earlier work, suggest that this part of *Starry Messenger* was largely composed prior to those observations, perhaps in September or October of that year.[62] But there are two aspects of *Starry Messenger* that seem to predate even the fall of 1609, and it is worth considering the possibility that they are remnants of Galileo's experimentation with a lens-and-mirror combination prior to June 1609.

One peculiar feature of *Starry Messenger* is its treatment of constellations. Galileo stated in his treatise that he had originally chosen to depict the entire constellation of Orion, and that lack of time and the sheer number of stars made visible by the telescope had forced him to postpone an undertaking of that size. Though Galileo's extant papers show no record of a map of the whole constellation, a letter from Sagredo suggests that perhaps this or another large constellation, prior to the discovery of the more spectacular satellites of Jupiter, figured as the original "Medici stars." Writing from Venice in 1612, Sagredo related, "I didn't observe the Medici Planets; being in Syria, I observed the Medici Stars with the first instrument that I had. Or rather, before I had it, I waited with great anticipation to observe the same constellations that you had. Then, upon reading *Starry Messenger,* I was amazed to have happened upon precisely the same part of the heavens. If you will send me your observations of the above-named Planets, it will be a reason for me to observe them."[63]

Sagredo appears to distinguish between the satellites and some constellations both he and Galileo observed, and his letter implies that one or more of the latter phenomena had originally been proposed as the "Medici stars." That he and Galileo turned to Orion does not seem surprising, as this large constellation would have been visible with and without the telescope from Padua and Aleppo in the winter and early spring; given that Sagredo expressed wonder at this coincidence, however, one might conclude that he knew no more than that his friend intended to dedicate some of the newly visible stars to the Medici.

It is possible this more modest plan dates to February 1609, the period in which Galileo was working on several undisclosed projects and seeking the attention of the new Grand Duke Cosimo II, and that his observations depended upon an instrument composed of a lens-and-mirror combination. Two anomalies in Galileo's eventual description of stellar observations in *Starry Messenger* suggest, at the very least, that he used a different and cruder instrument or method when examining the stars. First of all, his comparison of seventh-magnitude stars seen with the telescope and second-magnitude stars viewed with the naked eye is both inaccurate and difficult to reconcile with the small aperture on his Dutch telescope. Secondly, his failure to mention the nebula in Orion's sword, though this was quite possibly the effect of the haste with which this section of *Starry Messenger* was composed, might also indicate an earlier observation with an inferior instrument.[64]

The lens-and-mirror combination, however it was configured, would indeed have been inferior for celestial observations, and one of its very few associations with astronomy occurs in Thomas

Digges's 1576 edition of his father's *Prognostication Everlasting*. In this work, Digges portrayed and described an infinite universe thickly populated with stars, and his impression of the night sky has been attributed to his experiments with the instrument described five years earlier in the *Pantometria*. Because the foil backing of the concave mirror produced ghost images, what Digges saw was actually an overpopulated sky, a combination of newly revealed stars and illusions. In his view, these innumerable lights far excelled the sun both in quantity and quality.[65]

What Galileo had to say about the constellations is both much more detailed and balanced by the few observations we know to have been made with the Dutch telescope around February 1610, as the first part of his treatise was going to press. That said, a second peculiarity of *Starry Messenger* offers another residual trace of an earlier instrument and an earlier bid for the Grand Duke's favor. The eventual proem, addressed to Cosimo II de' Medici, announces the high point of the entire text, the recent discovery of Jupiter's satellites, but it starts out with an evocation of a distant and partially obliterated past, and specifically a past compatible with that of the Pharos. Paraphrasing a few lines from Roman poetry, Propertius's *Elegies* and Horace's *Odes,* Galileo states that those who seek to commemorate deeds and names rely upon images in marble and bronze, and upon equestrian and pedestrian statues, and that because of such desire for immortality, "the cost of columns and pyramids rises to the stars" and entire cities are named for those deserving of eternal memory.[66] He goes on to point out that those who sought out something yet more permanent entrusted celebration of their exploits to the "incorruptible monuments of letters," but concludes that these, too, eventually perish, and that the sole repository of eternity is the sky. But the stars, too, are sometimes vulnerable to change, Galileo warns, for when Augustus named a bright one after the late Julius Caesar, it soon disappeared, for it was no more than a comet.

Galileo soon ends his litany with the revelation about the four satellites meant to commemorate the Medici name; these, he says, as if in rejection of an earlier and less impressive scheme, are "not of the common sort and multitude of the less notable stars."[67] The most interesting feature of this otherwise melancholy exercise is the particular route it takes through a past portrayed as almost beyond recognition. Galileo evokes heroes whose names and images have endured an imperfect survival, a region that spared no expense for its columns and pyramids, a city named for a great individual, an institution where the "incorruptible monuments of letters" perished, and a conqueror whose bright bid for immortality was only a momentary flare. This past is not wholly obscured: it is clearly that of the Pharos, the monument variously associated with heroes whose exact names had vanished, the structure that perished while costly pyramids and columns endured, in the city named for Alexander the Great, alongside the library that was reportedly destroyed in a quick blaze set by Julius Caesar. The whole mechanism of collective memory, moreover, is portrayed in terms consonant with an optical device like the camera obscura, for Galileo affirms that "such is the condition of the human mind that unless continually struck by images of things rushing to it from the outside, all memories continually escape from it."[68]

What this set of ghost images in the text suggests, finally, is that Galileo might have originally intended to present one or both versions of the lens-and-mirror combinations as the glorious retrieval of the Pharos. It is much more likely that the treatise would have been a rotund evocation of the triumph of humanism and the ultimate accessibility of the past, and for practical reasons it would have insisted on the utility of the instrument in surveying the sea or the countryside rather than in observing the heavens. The details eventually supplied by Badovere, the superiority of the Dutch telescope, and above all the magnificent celestial discoveries it made available, would have all rendered the evocation of the Pharos useless, and nothing but the rubble of that rediscovery remains in the revised proem of *Starry Messenger*. But as the final chapter will show, the initial confusion of the real device with its mostly legendary predecessor persisted briefly and eventually became a sort of conflation, the Pharos offering a cultural template for the new invention.

Chapter Five

The Afterlife of a Legend

O F T H E many disputes surrounding Galileo, his alleged appropriation of the Dutch telescope is almost certainly the best known. This popular impression, which emerged in August 1609 in connection with della Porta's claim to priority, is somewhat surprising, given that Galileo never professed to have invented the telescope, and that, as this book has argued, his earliest concept of that instrument's design was inaccurate. The notion that led him, Sarpi, and others astray in their first encounter with the news from the Netherlands had a certain afterlife, for the two devices appear to have been confused elsewhere. And the features of a legendary mirror and an actual telescope were also conflated, even, or perhaps especially, by those perfectly capable of distinguishing them.

Magini and the Market for Mirrors

Posterity has reserved an unenviable perch for Giovanni Antonio Magini, who was in his day, until Galileo's rather sudden emergence, certainly Italy's best-known astronomer and astrologer. Our impressions of Magini today are colored by his perceived role as the most surreptitious of Galileo's many antagonists.[1] But by the mid-eighteenth century, even those who considered him apart from his professional rivalries emphasized something sad and superannuated about that figure who had loomed so large in early modern culture, as if to suggest that the enormous intellectual stature he had once enjoyed had been simply a matter of corpulence. Thus readers of Benjamin Martin's *Biographia Philosophica* of 1764 learned, for instance, that Magini foretold the rapid approach of his own death, not because of any astrological skill but simply because "he was very fat and burly."[2]

What is at stake here, however, is not the tragicomic portrait of Magini as a flabby follower of a more rigorous new science, but rather the way in which he, like Galileo, misunderstood the first news of the telescope. Comparison of their early and erroneous impressions of that device, in other words, allows us to establish an environment in which such mistakes were plausible and alternative histories of the invention might flourish.

Magini's original interest in mirrors lay in the practical problems of altimetry. In 1592 he published a short treatise on modes of measurement with quadrants, several of which involved the Euclidean precedent of a plane mirror; other sections of this work would find their way into Baldessar Capra's manual on the compass in 1607.[3] In 1602 Magini brought out a somewhat inaccurate version of Ettore Ausonio's *Theoretical Discourse on the Concave Spherical Mirror*, and in his *Brief Instruction on the Appearances and Marvelous Effects of the Concave Spherical Mirror* of 1611 he mentioned that he had already presented several such instruments, of notable size and quality, to "various Princes of Italy, Cardinals, and other Lords who found them agreeable," among them a nephew of the late Pope Innocent IX, the French queen Marie de Médicis, and the Holy Roman Emperor Rudolph II.[4]

Though a decade older than the Pisan scientist, Magini apparently figured as a kind of continuator of Galileo, rather than as someone who had been more or less displaced by him, and the longevity of his reputation had to do with the afterlife of the mirrors he made. Not long after Magini's death, for instance, in his *Anatomy of Melancholy,* Robert Burton ran through a brief genealogy of the Dutch telescope, naming as its principal developers those connected in the popular imagination with the deployment of concave mirrors: he mentioned the fifth-century philosopher Proclus, who was said to have imitated Archimedes' burning mirrors, and Roger Bacon's "burning glasses, multiplying glasses, perspectives," and he concluded by acknowledging that "glasses are much perfected of late by Baptista Porta and Galileo, and much more is promised by Maginus and Mydorgus," the latter a draughtsman associated with a very well-known compendium of optical and catoptrical tricks, and an expert producer of costly mirrors and lenses.[5] Magini's fame as a purveyor of concave mirrors appears to have survived him for at least a century, for an account of a burning glass published by the Royal Society in 1732 describes one in Lyon as even bigger and better than those originally owned by the scholar of Bologna, and apparently still available for inspection.[6]

Extant documents of the early seventeenth century also attest to the interest rulers had in acquiring large concave mirrors from Magini. One such testimony comes from Jean Baptiste Duval, an Arabist who would later become secretary to Marie de Médicis and who was in Italy from 1607 to 1610 in the service of the French ambassador to Venice, Jean Bochard de Champigny.[7] In the course of travels with Champigny's son in early October 1608, Duval met Magini, whom he described as a "very learned mathematician who has made himself known to the world because of his great labors and of the works he has brought to light."[8] Whether this last allusion concerns Magini's edition of Ausonio's work is not clear, but it was indeed a large concave mirror in the scholar's house that Duval found most memorable, noting that one effect was "to show whatever one had hidden beneath one's gown and cloak."

Magini told his guest that he was at work on another and yet larger concave mirror, and that he hoped to have finished it within three months. It is not surprising to find, therefore, that in mid-January 1609 the astronomer approached his patron, the Duke of Mantua, about both mirrors, whose "beautiful and most agreeable effects" allegedly surpassed those described in della Porta's *Natural Magic.*

> I want to tell your Highness that if he were thinking of offering a present to the Queen of France, sending one of these large mirrors would be most pleasing to her, and she would see it as worth at least 2000 [*scudi*], though it will cost you rather little. And I can tell your Highness this for certain, because I found it out when I was visited by the son of the French ambassador to the Venetian Republic on his way to Rome, who was accompanied by Signore

Jean Baptiste Duval, secretary to the Queen of France. Duval told me that the book dealer [Gaspare] Bindoni, having given a small spherical mirror to her Majesty—a mirror of the sort I once gave to your Highness—for [only] 400 [*scudi dal sole*] because when it was cast had come out poorly on one side, and was less rounded and beautiful, found that she took great pleasure in it, and this same secretary has translated the *Theoretical Discourse on the Concave Spherical Mirror,* which I published in Latin some years ago, into Italian for her, and he wanted to see some other catoptrical tricks in order to show them to the Queen. She has also tried in vain to have an artisan in Paris make a similar, though somewhat larger, mirror, but he did not succeed, and it broke.[9]

What is most significant about Magini's letter is its presentation of these large mirrors as desiderata of the French royal family. Some of this may be hyperbolic—for Magini clearly hoped to persuade his patron of the great appeal of such looking glasses to the queen—but it is interesting to compare her alleged taste for mirrors with her consort's supposed dependence upon that of the Jesuit Pierre Coton. Both reactions, moreover, contrast markedly with Henri's lukewarm reaction to the Dutch telescope when he was actually presented with the device by Pierre Jeannin in The Hague in early 1609.[10]

The implicit connection between the emergent telescope and the legendary concave mirror—a connection that would turn to competition once the two optical tools were sufficiently distinguished as an actual instrument involving glass lenses and a fabled device of greater cultural prestige and weaker powers of magnification—is especially apparent when we try to reconstruct the first reactions of potential patrons to these objects. Encounters with both instruments occurred in the wake of the Frankfurt Fair of fall 1608, as if in confirmation of an essay of 1574 celebrating that venue as the place where not just books but also "certain most ingenious machines, worthy of Archimedes himself," were to be found.[11] In the interest of convincing the Duke of Mantua that others desired his mirrors, Magini offered, in the same letter of January 14, 1609, an account that bizarrely parallels that of the Dutch telescope's diffusion.

> At present there is also the Archduke Albert [of Austria, ruler of the Spanish Netherlands], who has requested one of these mirrors of me. Signore Gastone Spinola, Count of Bruay, was the one who wrote already more than four months ago about it to the Jesuit Father Mazarino, who lives here [in Bologna] now, saying that he should get in contact with me. To this I responded that I had one of those first mirrors, which is like the one that Your Highness has, and that I was at the Archduke's service, and in the matter of the price would defer to him. But fortunately, the bookseller Bindoni having entered into all of this—like a man that is a little more eager than is good for him—on the occasion of going to the Frankfurt Fair, he also went to the Spanish Netherlands and showed one of the original mirrors to the Archduke and the Infanta. As Signor Gastone [Spinola] wrote to Father Mazarino, they took great pleasure in it, and he added that he did not believe that his Highness [the Archduke] should take it from that guy [Bindoni] because of the swollen price he asked—400 or rather 600 [*scudi*]—and also because he did not think it was fair, given that the offer was not mine.[12]

The overly enterprising bookseller, Gaspare Bindoni—already referred to as *nostro specchiaro,* "our mirror maker," by one of

Magini's friends in 1602, and a decade later a figure in the orbit of Giambattista della Porta—here occupied the same role as one of the earliest sellers of the Dutch telescope.[13] According to Simon Mayr, Baldessar Capra's former tutor and fellow plagiarist in the matter of Galileo's geometrical compass, in mid- to late September 1608 his patron encountered at the same fair in Frankfurt "a Dutchman who had invented an object by means of which the most distant objects might be seen as though quite near."[14] Mayr's patron examined the instrument, and though it seemed to him to work well enough, apart from a cracked lens, he found the price too high. Where the Dutchman went with his damaged and expensive instrument is not clear—for he cannot be identical with the several men in the United Provinces who sought patents for the same invention between September 25 and October 15, 1608—but his actions, or our impressions of them, differ little from those of Bindoni, who proceeded from the fair around September 22 to the Spanish Netherlands with a small but costly concave mirror that was not, in fact, really his to sell.[15]

As Magini's letter suggests, moreover, Gastone Spinola, a famous collector of mathematical instruments, had already sought to procure the same mirror for the Archduke and Infanta from him through their Jesuit contact in Bologna. Interestingly, in this account Bindoni would have approached the regents just days before Gastone's kinsman, the famous Marquis Ambrogio Spinola, arrived talking of *his* offering to Albert and Isabella, the Dutch telescope.[16] According to the papal nuncio to the Spanish Netherlands, Guido Bentivoglio, in early October 1608, the marquis spoke to the archduke in glowing terms about the military capabilities of the device he had just seen at The Hague, and they both "were most desirous to obtain such an instrument, and indeed it happened that one came into their hands, although not of such perfection as the one owned by Count Maurice [of Nassau]."[17]

Some of the structural similarities in these accounts of the two instruments are overdetermined: sellers of both the Dutch telescope and the concave mirror would naturally find the Frankfurt Fair the best venue and, in the absence of competitors, would inevitably seek inflated fees for their wares, while buyers would just as inevitably emphasize the relative inferiority and expense of the item at hand, and gesture toward better glasses elsewhere. There are significant differences, too, for in the case of the mirror, the unattainable superior product was the larger and later glass Magini had just finished in Bologna, whereas the first Dutch telescope was reportedly the best, and its immediate successor an imperfect copy. But what is more important, for our purposes, is the attempt that Magini made to present a story that is so similar, particularly in the chronology it offers, to contemporary accounts of a device that at once resembles and rivals his own. Magini's concern, moreover, was to establish some sort of priority: the rulers of France and the Spanish Netherlands had long been interested in his concave mirrors, and both parties had expressed particular curiosity about such items just as the Dutch telescope was emerging, and in the latter case, he alleged, shortly before the Frankfurt Fair.

One way of interpreting the rhetorical effort of Magini's letter is to suppose that he knew something in mid-January 1609 of the instrument lately developed at The Hague, and that he—like Galileo and Sarpi—also assumed at that point that it involved reflection rather than refraction. It was not, in other words, that he somehow sought to displace the lowly Dutch telescope with his more magnificent concave mirror, but rather that he believed one of its components to be the very item that he, rather than grasping booksellers or ambitious Dutch artisans, might best provide to desirous rulers. It is not clear that Magini knew precisely which optical qualities would be important in the telescopic mirror he imagined had just been perfected in the Netherlands. Apart from his insistence in this letter on the great size and weight of two mirrors in his possession, he mentioned that they were polished on both sides—offering thus convex and concave surfaces—and that the latter rendered inverted images and could ignite nearby targets. He concluded, however, that although the two mirrors "shared many effects, there is a notable difference between them, which means that it is worth the expense to have both of them," precisely for their unequal focal lengths and magnifying abilities.[18] It is possible, in other words, that Magini emphasized the range of catoptrical properties of his two mirrors because he hoped that one or the other—or best of all, both— would be useful in the production of the telescope.

If this was Magini's posture in January 1609, he shared with Galileo and Sarpi a preoccupation with the role of mirrors in optical devices. A little over a year later, when the *Starry Messenger* had emerged, matters were otherwise, for Magini suggested in a letter to a friend both that Galileo had been the dupe of an enormous optical illusion and that he was hoodwinking others with his device.[19] Galileo had occasion to experience Magini's hostility—constant but covert, and conducted by proxy—in late April 1609 when he stayed at the latter's home in Bologna for two nights and attempted to demonstrate the use of his telescope and to show some constellations and the satellites of Jupiter. The visit went quite poorly: neither Magini nor his colleagues were able to see any celestial object clearly, and Galileo's dual role as victim and perpetrator of a hoax appeared confirmed.

Most of what we know of this stay in Bologna comes from a singularly unpleasant source, Martin Horky. This Bohemian student lived in Magini's house as his scribe and wrote an attack on the *Starry Messenger* as well as a letter to Johannes Kepler in which he portrayed Galileo as a shuffling, gout-ridden syphilitic who was so shamed by his failures that he crept out of his host's house early one morning with his telescope, and without a word of thanks. Horky further related to Kepler—in a remark that undercuts his avowed contempt for the telescope—that he tried the instrument out repeatedly over the course of the visit. His letter, written in Latin, terminated with a German postscript where Horky confided that he had taken a wax impression of the instrument—presumably, of its lenses—and that, with God's help, upon his return to Prague he would be able to make a far superior telescope with this secret knowledge.[20]

Within about six weeks Horky saw into print his attack on Galileo's work, the *Brief Foray against the Starry Messenger,* and days later Magini turned his ungovernable scribe out of his house. At this point Horky found no better host than Galileo's old enemy Baldessar Capra, at whose home he hid most of the print run of

his *Brief Foray* while threatening to distribute the pamphlet unless repaid for his expenses.[21] Interestingly, Horky also made off
with some books pilfered from Magini's library, and it is clear
that of the texts he stole, the one that most interested him and
his former employer had to do with mirrors. In October 1610
Magini, anxious to dissociate himself from his scribe, and eager
to enlist Galileo's help in selling some of his large concave mirrors, wrote to the astronomer that he had verified that "that
rogue" had pilfered his library, and that among the most valuable
of his losses was a volume binding together four works. One had
to do with astrology, two with alchemy, and one was an early edition of Roger Bacon's *Letter on the Secret Works of Art and Nature.*
"In that book," Magini noted, "the author touched on a few fine
secrets about the concave mirror, saying that with it one would be
able to show [that is, project] on the moon an emblem to be understood by someone far away, but I don't remember if he recommended this so-called secret . . . Someone wrote me from
Modena that [Horky] was bragging about having this secret from
Bacon. I've verified that he robbed me of other books, ones which
don't matter much to me."[22]

Horky's interests, to judge from Magini's letter, were reminiscent of della Porta's in the 1580s when he first set out to work on
telescopic devices, for the Neapolitan philosopher had promised
Cardinal Luigi d'Este that he possessed the secrets that allowed
him to burn something a mile away or converse with a friend "by
means of the moon," and to make glasses that would show a man
a few miles away. Magini, for his part, would offer a scaled-down
and respectable version of the secret of satellite communication

in a publication of the following year, but what is most interesting here is Horky's apparent return to earlier discussions of the mirror, particularly as it would play out in the context of his *Brief Foray against the Starry Messenger.*[23]

A word is in order, finally, about Magini's adoption of the term *traguardo* for the concave lens used as an ocular in the Dutch telescope. As mentioned in Chapter 3, Magini took up the word once he had begun using the telescope, and it occurs occasionally in the following decades in this sense. The literal meaning of the word has a logic much better suited to the combination of lens and mirror than to the Dutch telescope, for the specific designation of the component through which one looked makes it clear that the concave mirror was the objective, and either a concave or a convex lens the ocular. Interestingly, *traguardo* appears to have been used to indicate the ocular *only* by those who, like Magini, studied the properties of large concave mirrors, and this trend sheds some retrospective light on Capra's obsession with the matter in his hapless appropriation of the geometrical compass. Thus Tiberio Spinola, a relative of Magini's associate Count Gaston Spinola, used it in a description of the Dutch telescope in 1621, and when he began discussing large concave mirrors some five years later with Cesare Marsili, the latter in his turn adopted the word and specifically indicated that it was the kind of lens that would be needed to be combined with such a mirror to achieve a telescopic effect. Marsili's next correspondent took up the word, agreeing that the traguardo would be the only way to bring out the mirror's telescopic features. He said he didn't have a concave mirror, and couldn't try it out, but he was clearly familiar enough with the argument. This was Galileo, writing in 1626, long after he, Sarpi, and Magini speculated about the device from The Hague. And the term was so well known as to be used without explanation in 1632 by Bonaventura Cavalieri in his work on burning mirrors, an important source of speculation about the device at the Pharos.[24]

Galileo in Egypt

In his *Brief Foray,* Horky followed Kepler's lead—to the latter's dismay—in asserting that della Porta had some claim to the invention of the telescope. Whereas Kepler had simply stated that the Neapolitan had "made public" the secret of the invention in his *Natural Magic,* Horky flatly denounced a theft with the sort of rhyming prose to which he was prone: "In the first place, [Galileo] took from Signor Giambattista della Porta, worthy in years and white hair, [*annis et canis honorandus*] what was his."[25] It turned out, however, that in Horky's analysis Galileo had borrowed from still older scholars. Horky reported, truthfully, that in April 1610 the astronomer had been unable to convince scholars from Magini's circle of the validity of his telescopic observations. He was especially dismissive of Galileo's specifications about the angular separations between Jupiter and its newly revealed moons; such claims were, he wrote, "one of the many false paradoxes, to use an Arabic or rather barbarian term," and the Jovian phenomena themselves merely the effect of a vapor-ridden atmosphere.

> This man has sold to all of you astronomers a manifest fiction, where he said that he had observed the new planets [that is, the

planetary moons] at so many degrees and minutes and seconds from Jupiter. Because he gave these angular separations, I, too, wanted to see these minutes and seconds with the same telescope, but I was unable to do this, because this glass has nothing that suffices even for the observation of degrees and minutes . . . Indeed, we are not with Galileo in Egypt [*non enim sumus cum Galilaeo in Aegypto*], where the sky is of perpetual serenity, but in Italy, where high mountains are near Padua, and against which the sun, moon, and other planets make various refractions. We were with Galileo near the Adriatic Sea, where there are denser vaporous exhalations, and thus greater refraction . . . Weep at the tomb of [the telescope's] stepmother, all you Galilean men, for though the discovery of astronomical instruments began many years ago, it is by no means finished.[26]

As foolish as Horky's remarks are, they are significant as a confused recycling of familiar cultural elements. In his account, Galileo is said to traffic in "Arabic or rather barbarian" specialties such as the paradox, a peculiar inference in that the paradox was of Greek origin: those of Zeno of Elea, transmitted in Aristotle's *Physics,* would be the most famous of the genre. What Horky might have had in mind, however, were popular collections of freakish natural facts, typically organized by geographical regions and known as "paradoxographia" or "wonders." Occupying a sort of crossroads of natural history, antiquarianism, travel literature, and romance, they had no necessary connection with Arabic culture but tended to involve foreign (that is, "barbarian") peoples and locales and were strongly associated with the Library of Alexandria, having first been compiled there by the scholar and poet Callimachus in the third century B.C. at the court of Ptolemy II.

Paradoxographia, in other words, would be the terrain in which many versions of the legend of the Pharos would flourish.[27]

It is notable, then, that Horky portrayed Galileo as once engaged in observing marvels in the serene skies of Egypt and failing to demonstrate the same to skeptics in Italy, and as an individual who had found or invented an imperfect astronomical instrument of a certain vintage. The astronomer occupied the dubious position of those traditionally associated with the Pharos, the vaguely identified "Ptolemy" and above all, the magus Apollonius of Tyana.[28] The terms of the argument likewise emerge from Pietro Pomponazzi's discussion of Apollonius's telescopic vision, for the implicit explanation appears to be a choice between the workings of a tricky Egyptian mirror and an effect that was the natural consequence of atmospheric conditions.

Horky's source for arguments about the nonexistence of the Jovian moons was in all likelihood an acquaintance in Florence, and he admonished Galileo to "listen to the young and most learned Francesco Sizi, a Florentine nobleman" and to read the letters of learned men sharing "their views of these four fake planets."[29] It is thus not surprising to find that when Sizi chose to follow Horky's disastrous lead in a publication of 1611, he attacked the validity of Galileo's observations of Jupiter's satellites by explicitly returning to the Egyptian lore. Recalling that Giambattista della Porta had written that "Ptolemy" had invented some sort of *perspicillum* and had placed it in the lighthouse of Alexandria, Sizi concluded that the ancient astronomer, so equipped, would have surely mentioned the moons of Jupiter if such really existed.[30]

Horky and Sizi had a similar goal—to discredit Galileo and his observations of Jupiter's moons—but they set about achieving it in slightly different ways. For Horky, Galileo was roughly comparable to a fraud like Apollonius of Tyana, for he claimed to have achieved an optical effect through a special instrument when atmospheric distortion was actually the agent of illusion. For Sizi, it was rather the legitimacy and priority of Ptolemy's telescope that made his silence about the satellites meaningful and undermined Galileo's apparent status as observer. Both insinuations had one final echo in the travel literature of the day, that of the contemporary naturalist Prospero Alpini.

Alpini took his medical degree at the University of Padua in 1578 and practiced there briefly before leaving for a trip to the Near East as physician to the Venetian consul in Cairo. He arrived in Alexandria in March 1581, some seven years before Hans Christoph Teufel's visit to that city, and returned in November 1584, shortly before the Austrian student arrived for his course of study in Italy. Though he published the medical and botanical works for which he is today remembered, *On Egyptian Medical Practice* and *The Plants of Egypt,* in the early 1590s, Alpini wrote up some of his first impressions of the Near East much later, between 1610 and 1616, and they emerged posthumously in his *Natural History of Egypt.* The first part of this work contains much in the way of ethnographic and cultural information and is organized according to the cities the naturalist visited. While the fourth chapter, devoted to Alexandria, makes no mention of the Pharos, the issue arises obliquely in a peculiar remark made pages earlier, in the general introduction, about ancient Egyptian astronomy. "The Egyptians, thanks to the purity and the clarity of the air, could with the greatest ease see and observe all the stars in the heavens, even the smallest ones, even though they did not have those spyglasses with a lens [*ocularia specillo crystallina*] only just lately invented, by which means Galileo, the Florentine Mathematician, observed and understood four new wandering stars, when he was at the famous University of Padua."[31]

It might be argued that the statement is designed chiefly to convey civic pride, and to remind a readership that lately associated Galileo solely with Tuscany that his earliest astronomical discoveries had in fact been made in Padua. But Alpini's explicit connection of Egyptian astronomy with talismans somewhat later in his *Natural History of Egypt* offers, at the very least, a context for the opposition of the legendary mirror to the actual telescope.[32] It is also probable that this particular passage was written in the wake of those two allusions to the fabled Egyptian device made by Horky and Sizi in 1610 and 1611.

Prospero Alpini's apparent need to insist that the ancient Egyptians had *not* had a telescope, and to point out that the device with lenses of rock crystal had only lately been invented, must therefore be understood in terms of these precedents, for it emerges from a context that had very recently, and erroneously, prized the Alexandrian mirror. What chiefly differentiates the observations of Teufel and Alpini, in other words, is hindsight: by the time the latter committed his early impressions of Alexandria to paper, the mythical Egyptian object was, for some readers, clearly outdistanced by the Dutch telescope and could no longer be proposed as its progenitor.[33]

Enduring Anachronisms

Although the familiar tradition of the telescopic mirror misled several of those who encountered the rumor about the Dutch telescope, it is also the case that those who were in a position to know the components of the new invention continued to associate it with the legendary device. Sometimes the pairing of the two instruments reflects a belief in their complementary nature and has to do with the rapid improvements the telescope, if not the lens-and-mirror combination, was then undergoing. Thus just two weeks after orchestrating the triumphal demonstration of the Dutch telescope during Galileo's visit to Rome in April 1611, the amateur scientist Federico Cesi was preoccupied above all with his studies of "mirrors and della Porta"; as a coded passage in his correspondence suggests, the Neapolitan's efforts initially appeared the latest counterpart, rather than the inadequate substitute, to those of the Pisan astronomer.[34] Elsewhere it appears that renewed insistence on the Pharos functioned as a way of casting doubt upon the claims of the *Starry Messenger* or shifting credit to della Porta. A typical and influential such instance is the Latin-language encyclopedia of Johann Heinrich Alsted, which neglected to mention either the telescope or Galileo's achievement in its survey of mathematical arts in 1613, but offered up instead a version of the Ptolemaic device, a concave mirror in a camera obscura through which the Egyptian ruler allegedly surveyed his distant troops.[35]

But the most curious examples are those that involve Galileo's deployment of a concave mirror, rather than a telescope, in the aftermath of the *Starry Messenger*. In two instances—presumably unrelated—he is presented not in the familiar guise of telescopic observer, but as one whose optical instruments have more immediate consequences: in an epic called *Florence Defended*, Galileo protects the Tuscan city with a burning mirror, while in Ben Jonson's satirical play *The Staple of News*, Galileo is in league with the Jesuits, who have taken over the moon, and he uses the same device to immolate ships at sea.[36] Stranger than these cases, and more closely associated with the tradition of the telescopic mirror, is an obituary poem published in 1613 in memory of Sir Thomas Bodley, founder of the Bodleian Library at Oxford. Written by Brian Twyne, a former student of the English mathematician and antiquary Thomas Allen, it offers what only now seems an embarrassingly anachronistic impression of the Dutch telescope:

To Galileo: Concerning Certain New Phenomena in the Moon

A number of things have been reported everywhere about that mirror you have, Paduan, and they concern your observations. And indeed I remember that you said a certain mountain protrudes beyond the lunar globe. An amazing belief! You were wrong: it is the shadow of this distinguished library, and if you'll believe me, of [Sir Thomas] Bodley's work. And soon we will project other shadows: towers and lecture halls, and things truly worthy of being seen in your mirror.[37]

The poem—apart from this graceless translation—is interesting, and it recapitulates many of the arguments of this book. Twyne knew full well the difference between the Dutch telescope and the telescopic mirror, but the latter seemed to him to evoke a richer tradition, and one that had a recognizable connection to English scholarship. As the student of Thomas Allen, he had had occasion to examine John Dee's great concave mirror, as well as the Baconian manuscripts held by Allen; those texts, in fact, would later be regarded by members of the Royal Society as crucial to Leonard and Thomas Digges's "revival" of the invention of the telescopic glass with the *Pantometria.* Although he was an enthusiastic follower of the *Starry Messenger,* Twyne proposed the storehouse of knowledge that the Bodleian represented as both the antecedent to Galileo's "mirror" and as a kind of future collaborator whose growth would be worth the attention of well-established observers on the Continent. Twyne's mistaken notion that Galileo was Paduan, rather than Pisan, may arise from the fact that so many English students had encountered him at the University of Padua, which was roughly equal in age and prestige to Oxford.[38]

It is notable, but not surprising, that in an obituary collection where Bodley was portrayed as the "Ptolemy of Oxford," Twyne made the shadows cast by the library and its adjoining buildings the phenomena under scrutiny in Galileo's "mirror." Much of what was proposed about telescopic mirrors prior to the emergence of the Dutch telescope was assumed to have been conditioned by the manner in which the relevant texts were variously acquired, preserved, hoarded, circulated, discussed, and even discarded. In their real and in their symbolic roles, libraries have figured throughout this narrative: the one at Alexandria appears as the replete storehouse and as the reminder of the transferability of technological secrets, that of the Vatican allowed Apollonius of Tyana to serve as a source of information on both conic figures and talismans, the manuscripts destined for the eventual Bodleian kept alive interest in Friar Bacon's glass, the collection at Gianvincenzo Pinelli's house in Padua offered limited access to the valuable document that was Ettore Ausonio's *Theoretical Discourse,* and the study in Giovanni Antonio Magini's house in Bologna encouraged the grasping Horky to seek out "a few fine secrets about the concave mirror," even in the wake of the invention in The Hague.

Brian Twyne's treatment of the curious temporality of inventions, finally, is nicely captured in his insistence upon Galileo's "mirror," at once a deliberate anachronism and the emblem of discoveries yet unmade, but more spectacular by far than that of the mountainous lunar surface. The belated and piecemeal European acquaintance with the myth of the Alexandrian mirror was a crucial factor in the plausibility it enjoyed in the latter part of the sixteenth century; though antecedent as a narrative to those involving Virgil, Prester John, and Friar Bacon, its relatively tardy appearance in Western culture made it persuasive in a way that its more familiar medieval analogues no longer were. "Why talk about publications and notoriety?" Galileo's spokesman Salviati would ask in the *Dialogue concerning the Two Chief World Systems.* "Does it make any difference whether the opinions and inventions are new to the people, or the people new to them?"[39] The fact that most early modern Europeans were relatively "new" to the old tale about the mirror at Alexandria meant that such a device could appear to at least some of them to be as novel, authentic, and appropriate for philosophical contemplation as the actual objects then manufactured, or rumored to be, closer to home. This comparison suggests, of course, a kind of fusion between stories that looked new and actual physical mirrors of recent, or rather, *future* construction, a curious collapsing of narrative and technical artifacts, as if the late discovery of an "opinion" about the legendary Pharos constituted a tangible and forthcoming "invention."

Notes · Acknowledgments · Index

Notes

Introduction: The Hague, 1608

1. On this transformation, see Mario Biagioli, *Galileo's Instruments of Credit: Telescopes, Images, Secrecy* (Chicago: University of Chicago Press, 2006), 10–12, 27–44, 77–134. For enlightening recent discussions of craft secrecy and humanist openness, and the relationship of artisanal knowledge to scientific practice, see Pamela O. Long, *Openness, Secrecy, Authorship: Technical Arts and the Culture of Knowledge from Antiquity to the Renaissance* (Baltimore: Johns Hopkins University Press, 2001), and Pamela H. Smith, *The Body of the Artisan: Art and Experience in the Scientific Revolution* (Chicago: University of Chicago Press, 2004).

2. Stillman Drake, "Galileo and the Telescope," in Drake, *Galileo Studies* (Ann Arbor: University of Michigan Press, 1970), 140–155; Drake, *Galileo at Work: His Scientific Biography* (Chicago: University of Chicago Press, 1978), 137–139; Gaetano Cozzi, *Paolo Sarpi tra Venezia e l'Europa* (Turin: Giulio Einaudi, 1979), 179–180; Galileo Galilei, *Sidereus Nuncius or The Sidereal Messenger,* trans. and annot. Albert van Helden (Chicago: University of Chicago Press, 1989), 4–5; Galileo Galilei, *Le Messager céleste,* trans. with introduc-

tion and notes by Isabelle Pantin (Paris: Belles Lettres, 1992), xiv–xix, 58–59; Galileo Galilei, *Sidereus Nuncius,* ed. Andrea Battistini, trans. Maria Timpanaro Cardini (Venice: Marsilio, 1993), 192–195; Michael Sharratt, *Galileo: Decisive Innovator* (Cambridge: Cambridge University Press, 1994), 13–15; James Reston Jr., *Galileo: A Life* (New York: HarperCollins, 1994), 85–88; Noel Swerdlow, "Galileo's Discoveries with the Telescope and Their Evidence for the Copernican Theory," in *The Cambridge Companion to Galileo,* ed. Peter Machamer (Cambridge: Cambridge University Press, 1998), 244–270, on 244–245.

3. For an excellent collection, discussion, and translation of the relevant documents, see Albert van Helden, *The Invention of the Telescope* (Philadelphia: American Philosophical Society, 1977), 35–36.

4. See Johannes Walchius, *Decas fabularum humanis generis* (Strasbourg: L. Zetzneri, 1609), 248; Galileo Galilei, *Sidereus Nuncius,* in *Le Opere,* ed. Antonio Favaro, 20 vols. (Florence: Giunti Barbèra, 1968), 3(1):60; Simon Marius, *Mundus iovialis* (Nuremberg: Johannes Lauer, 1614), pref.; Luigi Lollini, *Epistolae Miscellaneae* (Belluno: Francisco Vieceri, 1641), 212. The Dutch telescope is identified as Flemish in origin by Rafael Gualterotti in Galileo, *Opere,* 10:341, and by Alessandro Allegri, *Lettere di Ser Poi Pedante* (Bologna: Vittorio Benacci, 1613), 14, among others.

5. Van Helden, *Invention of the Telescope,* 37–40.

6. Ibid., 47–48. In *Sommaire description de la France, Allemagne, Italie, & Espagne* (Geneva: Iacob Stoer, 1605), 288, Théodore de Mayerne states that the Frankfurt Fair began on September 7 and lasted fifteen days.

7. Engel Sluiter, "The Telescope before Galileo," *Journal for the History of Astronomy* 28 (1997): 223–234; and J. C. Houzeau, "Le télé-scope à Bruxelles, au printemps de 1609," *Ciel et Terre* 3 (1882): 25–28.

8. [Great Britain, Royal Commission on Historical Manuscripts], *Papers of William the Trumbull the Elder,* ed. E. K. Purnell and

A. B. Hinds, 6 vols. (London: HMSO, 1924–1995), 2:90, 97, 104, 106; for more on the trade in early telescopes among English diplomatic personnel in the Spanish Netherlands, see 2:186, 228, 229, 238, 239, 268.

9. Van Helden, *Invention of the Telescope,* 41–42.

10. Ibid., 43; Pierre Jeannin, *Négociations,* in Claude-Bernard Petitot, ed., *Collection des mémoires relatifs à l'histoire de France,* 69 vols. (Paris: Foucault, 1820–1829), 5:82.

11. Van Helden, *Invention of the Telescope,* 44, 47–49; Galileo, *Opere,* 10:230, 252, 255; and Allan Chapman, "The Astronomical Work of Thomas Harriot (1560–1621)," *Quarterly Journal of the Royal Astronomical Society* 36 (1995): 97–107, on 101.

12. Van Helden, *Invention of the Telescope,* 39–40, 46–47.

13. Ibid., 48–51.

14. Ibid., 44–46; and Galileo, *Opere,* 10:252, 508, and 11:611–612.

15. Paolo Sarpi, *Lettere ai Gallicani,* ed. Boris Ulianich (Wiesbaden: Franz Steiner Verlag, 1961), 240; and Johannes Kepler, *Dissertatio cum Nuncio Sidereo,* in Galileo, *Opere,* 3(I):108, 109.

16. Galileo, *Opere,* 3(i):135–136, 158–159, 238–240, 329, and 10:390, 430.

17. Galileo, *Opere,* 13:57.

18. Pierre Gassendi, *Vita Peireskii,* in *Opera Omnia,* 6 vols. (Stuttgart: F. Fromann, 1964), 5:275.

19. Walchius, *Decas fabularum,* 247–248.

20. Franco Palladini, "Un trattato sulla costruzione del cannocchiale ai tempi di Galilei: Principi matematici e problemi tecnologici," *Nouvelles de la République des Lettres* 1 (1987): 83–102, on 102.

21. See the satirical discussion of the telescope in Francisco de Quevedo's "Holandeses en Chile," in *La Hora de todos y la Fortuna con seso,* ed. Jean Bourg, Pierre Dupont, and Pierre Genest (Madrid: Catedra, 1987), 310–313, esp. 313.

22. More generally, on the related assumption that the Dutch telescope and other modern optical instruments had antecedents in

antiquity, see Th. Henri Martin, "Sur des instruments d'optique faussement attribués aux anciens par quelques savants modernes," *Bullettino di bibliografia e di storia delle scienze matematiche e fisiche* 4 (1871): 165–238.

23. On the so-called Elizabethan telescopes, see A. J. Turner, "The Prehistory, Origins, and Development of the Reflecting Telescope," *Bollettino del Centro internazionale A. Beltrame di storia dello spazio e del tempo* 3–4 (1984): 11–22; Colin A. Ronan, "The Origins of the Reflecting Telescope," *Journal of the British Astronomical Association* 101, no. 6 (1991): 335–342; Colin A. Ronan, "Leonard and Thomas Digges," *Endeavour,* n.s., 16, no. 2 (1992): 91–94; Joachim Rienitz, "'Make Glasses to See the Moon Large': An Attempt to Outline the Early History of the Telescope," *Bulletin of the Scientific Instrument Society* 37 (1993): 7–9; Colin A. Ronan, "There Was an Elizabethan Telescope," *Bulletin of the Scientific Instrument Society* 37 (1993): 2–3; Gerard L'E. Turner, "There Was No Elizabethan Telescope," *Bulletin of the Scientific Instrument Society* 37 (1993): 3–5; Colin A. Ronan, "The Invention of the Reflecting Telescope," *Yearbook of Astronomy* (1993): 129–140; Ewen A. Whitaker, "The Digges-Bourne Telescope: An Alternative Possibility," *Journal of the British Astronomical Association* 103, no. 6 (1993): 310–312; Colin A. Ronan, "Postscript concerning Leonard and Thomas Digges and the Invention of the Telescope," *Endeavour,* n.s., 17, no. 4 (1993): 177–179. I am indebted to Albert van Helden and Noel Swerdlow for the first and third diagrams, and to figure 1 of Colin Ronan's "Postscript concerning Leonard and Thomas Digges" for the second one.

1. The Daily Mirror of Empire

1. References to the story occur in the *Romance of the Seven Sages,* in some of the Latin, Catalan, Italian, and English versions of this

work, in John Gower's *Confessio amantis,* in the *Chroniques de Tournay,* in the *Roman de Renart contrefait,* in the *Cleomadès,* in the *Chronique rimée* of Philippe Mousquet, and in the *Destruction de Rome.* On the complicated textual history of the *Romance of the Seven Sages,* see the introduction in Gaston Paris, ed., *Deux redactions du Roman des Sept Sages de Rome* (Paris: Firmin-Didot, 1876); Hans R. Runte, J. Keith Wikeley, and Anthony J. Farrell, *The Seven Sages of Rome and the Book of Sindbad: An Analytical Bibliography* (New York: Garland, 1984); and Hans R. Runte, "From the Vernacular to Latin and Back: The Case of *The Seven Sages of Rome,*" in Jeanette Beer, ed., *Medieval Translators and Their Craft* (Kalamazoo: Western Michigan University, 1989), 93–133, esp. 94–101.

2. *Le Roman de Renart contrefait,* ed. Gaston Raynaud and Henri Lemaître (Geneva: Slatkine, 1975), vv. 29391–29398, p. 71. All translations are mine, unless otherwise indicated. On the motif of Virgilian magic, particularly as it pertains to the imperial mirror, see *Early English Prose Romances,* ed. William J. Thoms, 2 vols. (London: Nattali and Bond, 1858), 2:3–17; A. Loiseleur Deslongchamps, *Essai sur les fables indiennes et sur leur introduction en Europe, suivi du Roman des Sept Sages de Rome en prose* (Paris: Techener, 1838), 150–155, 51–53; Édélestand du Méril, *Mélanges archéologiques et littéraires* (Paris: Franck, 1850), 437, 441, 447, 469–471; Arturo Graf, *Roma nella memoria e nelle imaginazioni del medioevo,* 2 vols. (Turin: Ermanno Loescher, 1882), 206–208; and Domenico Comparetti, *Vergil in the Middle Ages,* trans. E. F. M. Benecke (London: Swan Sonnenschein and Co., 1895), 303–305.

3. See Graf, *Roma nella memoria,* 207–208 n. 47; Du Méril, *Mélanges archéologiques,* 437 n. 2; Comparetti, *Vergil,* 304–305; Jean d'Outremeuse, *Le Myreur des Histors,* ed. A. Borgnet, 7 vols. (Brussels: M. Hayez, 1864–1887), 1:229.

4. On the Tower of Hercules at La Coruña, see Alexandre Haggerty

Krappe, "Une Légende de Coruña," *Bulletin hispanique* 33, no. 3 (1931): 193–198; and Krappe, "Studies on the *Seven Sages of Rome*," *Archivum romanicum* 16 (1932): 270–282, on 275–276.

5. Paulus Orosius, *Adversus paganos historiarum libri septem*, 1.20, in Jacques Paul Migne, ed., *Patrologiae cursus completus: Series Latina*, 221 vols. (Paris: Migne, 1844–1864), 31: col. 0688B.

6. Leomarte, "Como Girion fuyo a Galyzia e Hercules en pos del e lo alanço e lo mato," in *Sumas de Historia Troyana*, ed. Agapito Rey (Madrid: S. Aguirre, 1932), vol. 51, on p. 139; and Raoul Lefèvre, "Comment Herculés fonda la cité de La Courrongne sus la [tombe] de Gerion," in *Le Recoeil des Histoires de Troyes*, ed. Marc Aeschbach (Bern: Peter Lang, 1987), vol. 64, on pp. 388–389, 523–524. On Leomarte's pose as a chronicler, see Marina S. Brownlee, "The Trojan Palimpsest and Leomarte's Metacritical Forgery," *Modern Language Notes* 100, no. 2 (1985): 397–405.

7. See *The Recuyell of the Historyes of Troye*, ed. H. Oskar Sommer (London: David Nutt, 1894), bk. 2, chap. 22, on pp. 414–415.

8. *Epistola* (71) in Friedrich Zarncke, "Der Priester Johannes," *Abhandlungen d. phil-hist. Kl. D. Königl. Sächs. Ges. D. Wissenschaften* 7 (1879): 827–1030, on 920. For a general discussion of the relationship of the *Epistola* to scientific culture, see Lynn Thorndike, *A History of Magic and Experimental Science*, 8 vols. (New York: Macmillan, 1923–1958), 2:235–245. On the relationship of the Prester John myth to events in both Asia and Europe, see Francis M. Rogers, *The Quest for Eastern Christians: Travels and Rumor in the Age of Discovery* (Minneapolis: University of Minnesota Press, 1962); and J. R. S. Phillips, *The Medieval Expansion of Europe* (Oxford: Oxford University Press, 1988). On the relationship of the Latin *Epistola* to both preexistent legends and subsequent vernacular versions, see Vsevolod Slessarev, *Prester John: The Letter and the Legend* (Minneapolis: University of Minnesota Press,

1959). For useful compilations of different versions of the Latin and vernacular texts, see *The Hebrew Letters of Prester John,* ed. Edward Ullendorff and C. F. Beckingham (Oxford: Oxford University Press, 1982), 183ff.; and (with translations into Italian) *La lettera del Prete Gianni,* ed. Gioia Zaganelli (Parma: Prattiche Editirice, 1990).

9. *Epistola* (67–68) in Zarncke, "Der Priester Johannes," 919.

10. This is the poem of Osswalt der Schribar; see Zarncke, "Der Priester Johannes," 1015–1028.

11. See *La Lettre du Prêtre Jean: Les versions en ancien français et en ancien occitan—Textes et commentaires,* ed. Martin Gosman (Groningen: Bouma's Boekhuis, 1982), vv. 843–851, p. 136. The notion that the device's greatest potential was its ability to provide fresh news was still current, as a fictional gag, in a popular late seventeenth-century novel, Giovanni Paolo Marana's *Letters writ of a Turkish spy, who liv'd five and forty years undiscover'd at Paris;* in vol. 5, p. 51, the Turkish spy reporting on the French regime complains that the secretary of Nazarene affairs, in failing to keep him au courant of affairs in the Levant, must imagine that he can get by with a magical glass (Giovanni Paolo Marana, *The eight volumes of letters writ by a Turkish spy* [London: G. Strahan, 1730], 5:51).

12. The polished column that reveals images of each land as it revolves is described in strophe 590 in chapter 12 of the poem; for an English version, see Wolfram von Eschenbach, *Parzival,* trans. Arthur Thomas Hatto (London: Penguin, 1980), 297.

13. See Wolfram, *Parzival,* strophe 656, chap. 13; strophe 592, chap. 12; strophe 822, chap. 16; strophe 589, chap. 12—respectively pp. 328, 298, 408, and 297 in Hatto's translation.

14. For a convincing discussion of medieval magic as an occult and purposive rationality, distinct from everyday physical phenomena in that it depended upon demonic agents in the way that normal

operations of nature do not, see Richard Kieckhefer, "The Specific Rationality of Medieval Magic," *American Historical Review* 99, no. 3 (1994): 813–836.

15. On the question of sources for this tale, see John Livingston Lowes, "The Squire's Tale and the Land of Prester John," *Washington University Studies* 1, no. 2 (1913): 3–18; H. S. V. Jones, "The Squire's Tale," in *Sources and Analogues of Chaucer's Canterbury Tales,* ed. W. F. Bryan and Germaine Dempster (Chicago: University of Chicago Press, 1941), 357–369; and Helen Cooper, *Oxford Guides to Chaucer: The Canterbury Tales* (Oxford: Clarendon Press, 1989), 217–227.

16. "The Squire's Tale," vv. 82, 132–141, in *The Complete Poetry and Prose of Geoffrey Chaucer,* ed. John H. Fisher (New York: Holt, Rinehart and Winston, 1989), 189–190.

17. Ibid., vv. 221–224.

18. ibid., vv. 225–235.

19. Guillaume de Lorris and Jean de Meun, *Le roman de la rose,* ed. Ernest Langlois, 5 vols. (Paris: Firmin-Didot, E. Champion, 1914–1924), 5:213, 217–218, vv. 18044–18060, 18153–18179.

20. The distance is the maximum allowed, given the curvature of the earth; on this point, and on the structure of the Pharos, see Lucio Russo, *The Forgotten Revolution,* trans. Silvio Levy (Heidelberg: Springer, 2004), 115–118.

21. On the library, see Diana Delia, "From Romance to Rhetoric: The Alexandrian Library in Classical and Islamic Traditions," *American Historical Review* 97, no. 5 (1992): 1449–1467; and Daniel Heller-Roazen, "Tradition's Destruction: On the Library at Alexandria," *October* 100 (2002): 133–153.

22. On the fascination that Byzantium held for the Arab world in this period, see Nadia Maria El Cheikh, *Byzantium Viewed by the Arabs* (Cambridge, MA: Harvard University Press, 2004); on the ques-

tion of the Byzantines as "mere" craftsmen and as unworthy heirs to Greek science and philosophy, see esp. 100–111, 193–195; on the craftiness of the Byzantines and the weakness and political and cultural fragmentation of the Arabs, see 120–122, 175–177; on Constantinople, like a latter-day Alexandria, as an ageless city of marvels, see 139–152, 207–213. On Alexandria's commerce, particularly with Venice, in the medieval period and Renaissance, see Olivia Remie Constable, *Housing the Stranger in the Mediterranean World: Lodging, Trade, and Travel in Late Antiquity and the Middle Ages* (Cambridge: Cambridge University Press, 2003), 112–126, 136–137, 234–236, 243, 267–290; and Giovanni Curatola, "Venetian Merchants and Travellers in Alexandria," in Anthony Hirst and Michael Silk, eds., *Alexandria, Real and Imagined* (Aldershot: Ashgate, 2004), 185–198.

23. On accounts of the Pharos, see above all Hermann Thiersch, *Pharos: Antike Islam und Occident: Ein Beitrag zur Architekturgeschichte* (Leipzig: B. G. Teubner, 1909); and Miguel Asín Palacios, "Una descripción nueva del Faro de Alejandría," *Al-Andalus* 1 (1933): 241–292. As regards the issue of the way in which the lighthouse functioned, these authors differ primarily in their interpretation of the mirror; Thiersch believes some telescopic effect might have been achieved through combination with a camera obscura within the tower's interior (*Pharos*, 90–96), while Asín Palacios emphasizes the hyperbolic tone and marvelous intent of the passages mentioning the mirror. See also Ugo Monneret de Villard, "Il Faro di Alessandria secondo un testo e disegni arabi inediti da Codici Milanesi Ambrosiani," *Bulletin de la Société archéologique d'Alexandrie* 18 (1921): 13–35; Evariste Lévi-Provençal, "Une description arabe inédite du Phare d'Alexandrie," *Mémoires de l'Institut français d'archéologie orientale du Caire* 68 (1940): 161–171; François de Polignac, "*Al-Iskandariyya:* Œil du monde et

frontière de l'inconnu," *Ecole française de Rome: Mélanges—Moyen Âge—Temps Moderne* 96, no. 1 (1984): 425–439; and Faustina Doufikar-Aerts, "Alexander the Great and the Pharos of Alexandria in Arabic Literature," in *The Problematics of Power: Eastern and Western Representations of Alexander the Great,* ed. M. Bridges and J. Ch. Bürgel (New York: Peter Lang, 1996), 191–202. For a valuable collection of travelers' impressions of Alexandria, including many of the Pharos, arranged in chronological order and translated into modern French, see Oueded Sennoune, *La description d'Alexandrie à travers les récits des voyageurs* (Alexandria: Centre d'Études Alexandrines, 2004).

24. Dolors Bramon, ed., *El Mundo en el siglo XII: Estudio de la version castellana y del "Original" Árabe de una geografia universal—"El tratado de al-Zuhrī"* (Barcelona: Editorial AUSA, 1991), 81, n. 377.

25. Abū Hāmid Al-Garnātī, *Tuhfat Al-Albāb (El Regalo de los Espíritus),* ed. and trans. Ana Ramos (Madrid: Fuentes Arábico-Hispanas, 1990), 47–49; and *Al Mu'rib 'An Ba'D 'Aŷā' ib Al-Magrib (Elogio de algunas maravillas del Magrib),* ed. and trans. Ingrid Bejarano (Madrid: Fuentes Arábico-Hispanas, 1991), 155–159.

26. Benjamin of Tudela, *Itinerario (Sefer massa' ot),* ed. and trans. Giulio Busi (Rimini: Luisè, 1988), 78–79.

27. For these early sources, see Thiersch, *Pharos,* 42, 46, 49–51; and Asín Palacios, "Una descripción nueva," 248–249, 269–283; as well as Ibn Khurradâdhbih, Ibn al-Faqîh al-Hamadhâni, and Ibn Rustih, *Description du Maghreb et de l'Europe au IIIe–IXe siècle,* trans. and annotated by Hadj-sadok Mahammed (Algiers: Éditions Carbonel, 1949), 18–19, 62–63, 106–107; and Ibn al-Faqîh al-Hamadhâni, *Abrégé du livre des pays,* trans. Henri Massé (Damascus: Institut français de Damas, 1973), 88–89.

28. See Thiersch, *Pharos,* 46; Asín Palacios, "Una descripción nueva," 248, 270; Lévi Provençal, "Une description arabe," 167.

29. See Thiersch, *Pharos,* 46, 49, 51; Asín Palacios, "Una descripción nueva," 279, 282; Al-Garnāti, *Tuhfat,* 47. On the relatively undeveloped glass industry in China, and the preference for bronze mirrors, see Alan MacFarlane and Gerry Martin, *Glass: A World History* (Chicago: University of Chicago Press, 2002), 109–112. On the Arabs' impression of the superiority of Chinese craftsmanship to that of the skilled Byzantines, see El Cheikh, *Byzantium Viewed by the Arabs,* 60; on the Arab reaction to the Mongol invasions, see 180–181, 191.

30. See Thiersch, *Pharos,* 46, 49; Asín Palacios, "Una descripción nueva," 278; Al-Garnātī, *Tuhfat,* 48; and Bramon, *El Mundo en el Siglo XII,* 81–82. Where the mirror is composed "of Chinese iron," it no longer functions after the collapse of the tower because of rust rather than breakage.

31. Although it is conceivable, for instance, that the Liegeois chronicler Jean d'Outremeuse's unusual emphasis on the maritime uses of the Roman device is due to the influence of the Alexandrian model, it is more likely that his model was that of La Coruña. See d'Outremeuse, *Le Myreur des Histors,* 1:229.

32. On the manuscripts and publication history of Benjamin's travel narrative, see Guilio Busi's edition of the *Itinerary,* 85–87.

33. Cited in Asín Palacios, "Una descripción nueva," 280.

34. On the interest in Apollonius of Perga, particularly in Padua in the early 1580s, see Carlo Maccagni and Giovanna Derenzini, "Libri Apollonii qui . . . desiderantur," in *Scienza e filosofia: Saggi in onore di Ludovico Geymonat,* ed. Corrado Mangione (Milan: Garzanti, 1985), 678–696; and Robert B. Todd, "Henry and Thomas Savile in Italy," *Bibliothèque d'humanisme et Renaissance* 58, no. 2 (1996): 439–444.

35. On the association of Apollonius of Tyana with talismans among Middle Eastern writers—a connection that Philostratus sought to

undercut in his *Life of Apollonius of Tyana*—see *Vita di Apollonio di Tiana,* trans. Dario del Corno (Milan: Adelphi, 1978), 43–44. For early modern European allusions to Apollonius's use of talismans, see Johannes Hottinger, *Historia orientalis* (Zurich: Bodmer, 1660), 286–287; and M. Le Nain de Tillemont, *An Account of the Life of Apollonius Tyaneus* (London: Printed for S. Smith and B. Walford, 1702), 21–22, 61. For Apollonius of Tyana's importance in Byzantium, see El Cheikh, *Byzantium Viewed by the Arabs,* 147–149, 209.

36. On the confusion of the two men, separated by three centuries but both known as "Balīnūs," among Arabic writers, see M. Plessner, "Balīnūs," *Encyclopedia of Islam,* ed. H. A. R. Gibb, E. Lévi-Provençal, and J. Schacht, 9 vols. (Leiden: E. J. Brill; London: Luzac and Co., 1960), 1:994–995. For Cassiodorus's reference, see *M. Aurelii Cassiodori chronicon, Ad Theodorum Regem* in Migne, *Patrologiae cursus completus: Series Latina,* vol. 69, col. 1231B. Naudé's remarks occur in Gabriel Naudé, *Apologie pour tous les grands personnages qui ont esté faussement soupçonnez de magie* (The Hague: Adrian Vlac, 1653), 292–293. Naudé's reference to Boissard concerns the *Tractatus posthumus Jani Jacobi Boissard Vesuntini, De divinatione magiciis præstigiis* (Oppenheim: Hieronymus Galler, 1617). On p. 343 Boissard stated that Apollonius of Tyana "was a celebrated mathematician, and among other works of his is the noble book *On Conics,* which Rafael Maffei of Volterra said that he used in Rome in the Vatican Library, in bk. 13, chap. 3, of his *Commentariorum Urbanorum.* Philostratus also affirmed that the same man wrote four books of astrological divination." Although Maffei did identify Apollonius of Perga as an *astrologus* "who among other things wrote *On Conics,* a noble treatise that can be seen in the Vatican Library," he appears to have distinguished him, in a way that Boissard certainly did not, from Apollonius of Tyana, whose activities he subsequently discussed at greater length. See

Commentariorum Urbanorum Raphaelis Volaterrani (Rome: n.p.,
1506), fols. 183–185v. Boissard's effort is to associate Apollonius of
Tyana with the lighthouse at Alexandria; on page 338 of the
Tractatus, Apollonius's teacher, the Indian magus Iarchus, is sugges-
tively identified as a former ruler of Egypt and an inhabitant of the
Pharos. Boissard's remarks also appear in the third book of the alchemist Michael Maier's *Symbola aureae mensae duodecom nationum*
(Frankfurt: Antonius Hummius, 1617), 126. Finally, Naudé's reference to Pierre de L'Ancre concerns his *L'incrédulité et mescréance du
sortilege plainement convaincue* (Paris: N. Buon, 1622), 60, a discussion of Apollonius of Tyana's sorcery that includes the misunderstood passage from Maffei.

37. For other references to the talismanic power of the mirror, see
Thiersch, *Pharos,* 42, 46; and Asín Palacios, "Una descripción
nueva," 273.

38. Roger Bacon, *Frier Bacon His Discovery of the Miracles of Art, Nature, and Magick, faithefully translated out of Dr Dees own Copy, by
T. M. and never before in English* (London: Printed for Simon
Miller, 1659), 20–21; cited in part in Albert van Helden, *The Invention of the Telescope* (Philadelphia: American Philosophical Society,
1977), 28.

39. On Caesar's alleged construction of the Pharos, see, for instance,
the late twelfth-century account offered by William of Tyre in
Sennoune, *Description d'Alexandrie,* 42. On the varying accounts of
the immolation of the library, see Delia, "From Romance to Rhetoric," 1460–1462; and Heller-Roazen, "Tradition's Destruction,"
147–151.

40. A. G. Molland, "Roger Bacon as Magician," *Traditio* 30 (1974):
445–460, on 454. Edward Rosen is less persuaded of this possibility; see his "Did Roger Bacon Invent Eyeglasses?" *Archives internationales d'histoire des sciences* 7 (1954): 3–15, on 14.

41. Roger Bacon, *The Opus Majus of Roger Bacon,* trans. Robert Belle Burke, 2 vols. (Philadelphia: University of Pennsylvania Press, 1928), 2:582; cited in Van Helden, *Invention of the Telescope,* 28.

42. Rosen, "Did Roger Bacon Invent Eyeglasses?" 8–10. For other classical and patristic references to water-filled globes allegedly used for combustion or for magnification, see Dimitris Plantzos, "Crystals and Lenses in the Graeco-Roman World," *American Journal of Archaeology* 1010 (1997): 451–464, on 459.

43. Molland, "Roger Bacon as Magician," 446–447, 453; Robert Recorde, *The pathway to knowledge, containing the first principles of geometrie* (London, 1551), pref.; cited in Van Helden, *Invention of the Telescope,* 29.

44. *The Famous Historie of Fryer Bacon* (n.p., 189?), 63.

45. For a typical instance, see George Adams, *Lectures on natural and experimental philosophy,* 5 vols. (London: R. Hindmarsh, 1799), 2:483–484.

46. On the *Oration,* see the valuable summary and discussion in Noel Swerdlow, "Science and Humanism in the Renaissance: Regiomontanus's *Oration on the Dignity and Utility of the Mathematical Sciences,*" in *World Changes: Thomas Kuhn and the Nature of Science,* ed. Paul Horwich (Cambridge, MA: MIT Press, 1993), 131–168; on the "philosophic mirror" in particular, see 147, 148, 161.

47. Pietro Pomponazzi, *De incantationibus liber,* 1:4, in *Opera* (Basel: Henricpetri, 1567), 57–58. For an important discussion of Pomponazzi's place in early modern science, see Brian P. Copenhaver, "Did Science Have a Renaissance?" *Isis* 83 (1992): 387–407.

48. *Nicolai Tignosii Fulignatis in libros Aristotelis de Anima commentarii* (Florence: Ex Bibliotheca Medicea, 1551), 176–177. The explanation is reiterated in Martin Del Rio SJ, *Disquisitionum magicarum libri sex* (Mainz: Johann Weiss, 1612), fol. 120v.

49. Jacques Gaffarel, *Curiositez inouyes sur la sculpture talismanique, des*

Persans (Paris?, 1650), 252. The first edition of this work appeared in 1629.

50. Girolamo Fracastoro, *Homocentrica,* 1:8 and 3:23, in *Opera Omnia* (Venice: Giunti, 1574), fols. 13r–v and 42.

51. Ibid., 1:8, in *Opera Omnia,* fol. 13v.

52. Ibid., 3:23, in *Opera Omnia,* fol. 42v.

53. A strong convex lens held against the eye would not change the size of the object; moved some inches away from the observer, the lens would yield first an erect image increasing in size but also in blurriness, and then an inverted image whose sharpness grows only as it diminishes. A combination of a strong and a weak convex lens would yield a sharp, slightly enlarged, and inverted image, but only if the lenses were separated by twenty-four inches, an arrangement that would undercut Fracastoro's insistence on the density of the medium as a corollary of magnification. See in this connection Van Helden, *Invention of the Telescope,* 17, 28.

54. See Galileo, *Opere,* 3 (I):329 and 13:57; Abraham Hill's statement to the Royal Society in 1682 (Thomas Birch, *The History of the Royal Society,* 4 vols. [London: A. Millar, 1756–1757], 4:157); and John Aikin, "Fracastoro," in *General Biography,* 10 vols. (London: for G.G. and J. Robinson, 1799–1815), 4:198.

55. *Hieronymi Fracastorii vita,* in *Opera Omnia,* fols. 3r–3v.

56. Ibid., fol. 3.

57. Leo Africanus, *Descrizione dell'Africa,* in Giovanni Battista Ramusio, *Navigazioni e Viaggi,* ed. Marica Milanesi, 6 vols. (Turin: Giulio Einaudi, 1978–1988), 1:397. On Leo Africanus, see Natalie Zemon Davis, *Trickster Travels: A Sixteenth-Century Muslim between Two Worlds* (New York: Hill and Wang, 2006).

58. See Jean-Léon L'Africain, *Description de l'Afrique,* new edition trans. from the Italian by Alexis Epaulard, 2 vols. (Paris: Librairie d'Amérique et d'Orient Adrien-Maisonneuve, 1956), 2:497. For the

story of the manuscript's discovery and for discussion of the problems with Épulard's editorial practices, see Davis, *Trickster Travels,* 5–7.

59. Johannes Trithemius, *Primae partis opera historica,* 2 vols. (Frankfurt: Wechel, 1601), 2:412. A similar story also appeared in Laurentius Surius, *Commentarius brevis rerum in orbe gestarum ab anno salutis M.D. usque in annum MDLXVIII* (Cologne: Gervinus Calenius and the heirs of Johann Quentels, 1568), 21–22. Both are mentioned in Thorndike, *Magic and Experimental Science,* 4:557–558.

60. For other discussions of Giovanni Mercurio, see David B. Ruderman, "Giovanni Mercurio da Correggio's Appearance in Italy as Seen through the Eyes of an Italian Jew," *Renaissance Quarterly* 28 (1974): 309–322.

61. So, too, did Henri's son, Louis XIII (1601–1643), according to Galileo's correspondent Cesare Marsili, who wrote in 1626 that a concave mirror "with telescopic effect" had been presented by Cesare Caravaggi and his assistant and heir Giovanni to the king of France. For discussion of this and similar mirrors, see Galileo, *Opere,* 13:314–315, 330–333, 335. Caravaggi is also mentioned by Giuseppe Biancani SJ, in his *Echometria* (Bologna: Girolamo Tamburini, 1620), 232.

62. Rafael Mirami, *Compendiosa introduttione alla prima parte della specularia* (Ferrara: Heirs of F. Rossi and P. Tortorino, 1582), 4.

63. See Vannoccio Biringuccio, *De la pirotechnia* (Venice: Venturino Rossinello, 1540), fol. 143r. I thank Sylvie Deswarte-Rosa for this reference. Biringuccio's story might have been associated with relatively novel glass mirrors of German production, mentioned at the closing of the chapter on fol. 144v.

64. Tomaso Garzoni, "De' Speculari, et Specchiari" (Discorso 145), in *La piazza universale di tutte le professioni del mondo,* ed. Giovanni

Battista Bronzino, Pina de Meo, and Luciano Carcereri, 2 vols. (Florence: Leo S. Olschki, 1996), 2:1090.

65. *The Bondage and Travels of Johann Schiltberger, a native of Bavaria, in Europe, Asia, and Africa, 1396–1427,* trans. J. Buchan Telfer, annotated by P. Bruun (London: Hakluyt Society, 1879), 62–64, 214–216.

66. Johannes Kepler, *Gesammelte Werke,* ed. Walther von Dyck and Max Caspar, 21 vols.(Munich: C. H. Beck'sche, 1937–), 16:317.

67. On Teufel's studies and career, see the brief remarks that precede the modern and abridged version of his travel narrative, *Voyages en Egypte pendant les années 1587–1588,* ed. Serge Sauneron, trans. from the Italian by Nadine Sauneron (Cairo: Institut français d'archéologie orientale, 1972), vi–vii. See also *Beiträge zur Geschichte der Niederösterreichischen Statthalterei* (Vienna: Friedrich Jasper, 1897), 242–243.

68. *Il viaggio del molto illustre signor: Giovanni Christophoro Taifel* (Vienna: Franz Kolb, 1598), chap. 2, cited in Sauneron, *Voyages en Egypte,* 147–148.

69. Benjamin appears in most versions of his *Itinerary* to have referred to a single mirror. The exception is the recent French translation prepared by Haïm Harboun, *Les Voyageurs juifs du XIIe siècle: Benjamin de Tudèle* (Aix-en-Provence: Éditions Massoreth, 1998), 281–282.

70. See Jacques Besson, *Théâtre des instrumens mathématiques & méchaniques,* ed. François Béroalde de Verville (Lyon: Barthélémy Vincent, 1578), prop. 42. Besson's earlier edition of this work, unavailable to me, was published around 1571 or 1572; as a Huguenot, Besson fled France for England in 1572 and may have died there a year later. He appears to have had among his contacts in France the English translator of technical treatises Richard Eden; see Eden's "Epistle Dedicatorie to the right woorshipfull Syr Wylliam

Wynter" in Joannes Taisnierus, *A very necessarie and profitable Booke concerning Navigation,* trans. Richard Eden (London: Richard Iugge, 1575), fols. 2v–3. On Besson's work, see also Alexander G. Keller, "The Missing Years of Jacques Besson, Inventor of Machines, Teacher of Mathematics, Distiller of Oils, and Huguenot Pastor," *Technology and Culture* 14, no. 1 (1973): 28–39; Keller, "A Manuscript Version of Jacques Besson's Book of Machines, with His Unpublished Principles of Mechanics," in *On Pre-Modern Technology and Science: A Volume of Studies in Honor of Lynn White, Jr.,* ed. Bert S. Hall and Delno C. West (Malibu, CA: Undena, 1976), 75–103; D. Hillard, "Jacques Besson et son 'Théâtre des instruments mathématiques,'" *Revue française d'histoire du livre* 22 (1979): 5–32; and Henry Heller, *Labour, science, and technology in France 1500–1620* (Cambridge: Cambridge University Press, 1996), 105–111.

71. Bonaventura Cavalieri, *Lo specchio ustorio* (Bologna: Clemente Ferroni, 1632), 126–128. On Cavalieri's subsequent suggestion, a combination of concave mirror, flat mirror, and concave glass ocular anticipating the Newtonian telescope, see Piero E. Ariotti, "Bonaventura Cavalieri, Marin Mersenne, and the Reflecting Telescope," *Isis* 66 (1975): 303–321, esp. 314–316.

72. See Cristoforo Armeno, "Prologo," in *Peregrinaggio di tre giovani, figliuoli del rè di Serendippo,* ed. Renzo Bragantini (Rome: Salerno Editrice, 2000), 33. On Cristoforo, his translation, and the influence of this work, see E. Melfi, "Cristoforo Armeno," *Dizionario biografico degli italiani,* ed. Alberto M. Ghisalberti, 65 vols. to date (Rome: Istituto della Enciclopedia Italiana, 1960–), 31:71–72.

73. See François Béroalde de Verville, *L'Histoire veritable, ou Le Voyage des princes fortunez* (Paris: P. Chevalier, 1610), 303. On the relationship of Béroalde's scientific and literary interests, see Stephen Bamforth, "Béroalde de Verville and the Question of Scientific Poetry," *Renaissance and Modern Studies* 23 (1979): 104–127;

Bamforth, "Béroalde de Verville and *Les Apprehensions Spirituelles,*" *Bibliothèque d'humanisme et Renaissance* 56, no. 1 (1994): 89–97; Bamforth, "Béroalde de Verville: Poète de la Connaissance," *Nouvelle Revue du Seizième Siècle* 14, no. 1 (1996): 43–55; and Neil Kenny, *The Palace of Secrets: Béroalde de Verville and Renaissance Conceptions of Knowledge* (Oxford: Clarendon Press, 1991). For the relationship of *The Travels of Three Young Men, Sons of the King of Serendippus* to *The Journey of the Fortunate Princes,* see Kenny, *The Palace of Secrets,* 191–203.

74. In addition to the interest in catoptrics in Padua, Teufel could have encountered Giovanni Antonio Magini at the University of Bologna and, in 1585–1586, Galileo at the University of Siena.

75. Georg Christoph Fernberger, *Reisetagebuch (1588–1593),* ed. and trans. from Latin to German by Ronald Burger and Robert Wallisch (Frankfurt: Peter Lang, 1999), 17.

76. Martin Crusius, *Turco Graeciae libri octo* (Basel: Leonhart Ostein, 1580), 231. In Benjamin's account, the crafty Greek is named "Theodorus" rather than "Sodorus."

77. Reinhold Lubenau, *Reisen,* in Sauneron, *Voyages en Egypte pendant les années 1587–1588,* 204.

78. Nikolaus Christoph Radziwill, *Peregrinatio* (Antwerp: Plantin, 1614), 198.

79. Cited in Sennoune, *Description d'Alexandrie,* 353.

80. Guillaume Bouchet, *Sérrés* (Lyon: Thibaud Ancelin, 1608), vol. 2, fols. 148v–149r. The second book of *Sérrés* first appeared in Paris in 1597.

2. Idle Inventions

1. On the "reading stone," or beryl, see G. R. Cardoni, "I nomi del berillo," *Incontri linguistici* 6 (1980): 63–96, esp. 70–77; and Franz Daxecker and Annemarie Broucek, "Eine Darstellung der hl.

Ottilie mit Lesensteinen," *Gesnerus* 52 (1995): 319–322. For discussion of the iconographic representation of other types of lens for presbyopes, see Franz Daxecker, "Three Reading Aids Painted by Tomaso da Modena in the Chapter House of San Nicolò Monastery in Treviso, Italy," *Documenta Ophthalmologica* 99 (1999): 219–223. On very early eyeglasses, see Judith S. Neaman, "The Mystery of the Ghent Bird and the Invention of Spectacles," *Viator* 24 (1993): 189–214; on later developments, especially in Florence, see Vincent Ilardi, "Eyeglasses and Concave Lenses in Fifteenth-Century Florence and Milan: New Documents," *Renaissance Quarterly* 29, no. 3 (1976): 341–360; Ilardi, "The Role of Florence in the Development and Commerce of Spectacles," *Atti della Fondazione Giorgio Ronchi* 56 (2001): 163–176; and Ilardi, "Renaissance Florence: The Optical Capital of the World," *Journal of European Economic History* 22, no. 3 (1993): 507–542. On the connection between the incidence of myopia and the growth of literacy, see Albert van Helden, *The Invention of the Telescope* (Philadelphia: American Philosophical Society, 1977), 10–11; Alan MacFarlane and Gerry Martin, *Glass: A World History* (Chicago: University of Chicago Press, 2002), 144–174; and, more generally, John Dreyfus, "The Invention of Spectacles and the Advent of Printing," *Library,* ser. 6, vol. 10, no. 2 (1988): 93–106.

2. *Le Roman de la Rose,* ed. Ernest Langlois, 5 vols. (Paris: Firmin-Didot, 1914–1924), 4:213, vv. 18048–18060.

3. Giovanni Rucellai, *Le Api,* vv. 972–985, in *Opere,* ed. Guido Mazzoni (Bologna: Nicola Zanichelli, 1887), 36–37.

4. "How Eulenspiegel became an optician, and found no demand for his services in any country," in *Till Eulenspiegel: His Adventures,* ed. and trans. Paul Oppenheimer (New York: Routledge, 2001), 126–128; on the foolishness of eyeglass consumers, see Jean-Claude Margolin, "Des lunettes et des hommes ou la satire des mal-

voyants au XVième siècle," *Annales: Economies, Sociétés, Civilisations* 30, no. 2 (1975): 375–393.

5. *Correspondance politique de MM. de Castillon et de Marillac, ambassadeurs de France en Angleterre (1537–1542),* ed. Jean Kaulek (Paris: Félix Alcan, 1885), 289. All translations are mine, unless otherwise indicated. I thank Sven Dupré for this reference.

6. Vannoccio Biringuccio, *De la pirotechnia* (Venice: Venturino Rossinello, 1540), fol. 143r.

7. Lionardo Fioravanti, *Dello specchio di scientia universale* (Venice: Zattoni, 1678), 93–94. In this edition, Ettore Ausonio's name is written "Ettor Eusobio." In the French translation of 1586, he is misidentified as "ce grand philosophe & Mathematicien Ensonio;" see *Miroir universel des arts et sciences* (Paris: Pierre Cavellat, 1586), 105.

8. Marin Mersenne, *Correspondance,* ed. Marie Prisset Tannery, Cornelis de Waard, and René Pintard, 17 vols. (Paris: Beauchesne, 1933), 8:211.

9. For the text, a diagram, and a brief discussion, see Sven Dupré, *Galileo, the Telescope, and the Science of Optics in the Sixteenth Century* (doctoral thesis, University of Ghent, 2002), 235–236, 325. I am very grateful to Professor Dupré for alerting me to this and subsequent passages from Ausonio's papers.

10. Girolamo Cardano, *De subtilitate* (Lyon: Etienne Michel, 1580), 168–169.

11. Girolamo Cardano, *De subtilitate* (bk. 4) in *Opera, tomus tertius quo continentur physica* (Lyon: Jean Antoine Huguetan and Marc Antoine Ravaud, 1663), 426; Giambattista della Porta, *Dei miracoli et maravigliosi effetti dalla natura prodotti libri IIII,* 4:9 (Venice: Lodovico Avanzi, 1560), 145v–146r; Rafael Mirami, *Compendiosa introduttione alla prima parte della specularia* (Ferrara: Heirs of F. Rossi and P. Tortorino, 1582), 41–42. Mirami's suggestion that

some of the mirrors had also to be magnifiers for the effect to be telescopic may derive from a device featured four years earlier in Jacques Besson's *Théâtre des instrumens mathématiques & méchaniques,* ed. François Béroalde de Verville (Lyon: Barthélémy Vincent, 1578): there, letters on a printed page, reflected in a plane mirror, were then "augmented in size by a concave one . . . with great benefit to the eyesight" (prop. 42).

12. On Coccio, see Paolo Procaccioli, "Note e testi per Francesco Angelo Coccio," *La Cultura* 27, no. 2 (1989): 387–417; for the text of the letter, see 416–417. For discussion, see Dupré, *Galileo,* 247–248.

13. Typical examples are *Epistulae ex Ponto,* 1.8.31–39, and *Tristitia,* 3.1.19–22, 27–36.

14. The best representative of Flemish landscape painting then in Venice was Lambert Sustris. On the importance of one of Coccio's translations for several of Sustris's paintings, see Michel Hochmann, *Peintres et commanditaires à Venise, 1540–1628* (Rome: École française de Rome, 1992), 152–153.

15. On Salviati's scientific interests, see Bruce Boucher, "Giuseppe Salviati, Pittore e matematico," *Arte Veneta* 30 (1976): 219–224; for his illustrations of Barbaro's work, see *I dieci libri dell'Architettura . . . tradutti e commentati da Monsignor Barbaro* (Venice: Francesco Marcolini, 1556). On the strong likelihood that Cesare Cesariano's 1521 commentary on Vitruvius, one of the earliest to describe the camera obscura and subsequently repeated by Giambattista Caporali in his translation of *De architectura* (Perugia, 1536), was the source of Daniele Barbaro's explicit adaptation of a biconvex lens to the aperture in his *Pratica della perspettiva* of 1568, see A. K. Wheelock, "Constantijn Huygens and Early Attitudes toward the Camera Obscura," *History of Photography* 1, no. 2 (1977): 93–103, on 99.

16. On the statue, see Massimiliano Rossi, *La poesia scolpita: Danese Cataneo nella Venezia del Cinquecento* (Lucca: Maria Pacini Fazzi, 1995), 67–75; on the influence of Salviati, see esp. 143–152, 184.

17. Cardano, *De subtilitate,* 169.

18. Ausonio, *Di una nuova invenzione d'uno specchio,* cited in Dupré, *Galileo,* 324.

19. Ibid.; and Cardano, *De subtilitate,* 169.

20. Ausonio, *Di una nuova invenzione d'uno specchio,* 323.

21. On Ausonio's *Theorica,* see above all Sven Dupré, "Mathematical Instruments and the 'Theory of the Concave Spherical Mirror': Galileo's Optics beyond Art and Science," *Nuncius* 15 (2000): 551–588; and Dupré, "Ausonio's Mirrors and Galileo's Lenses: The Telescope and Sixteenth-Century Practical Optical Knowledge," *Galilaeana* 2 (2005): 145–180. More generally see Pasquale Ventrice, "Ettore Ausonio matematico dell'Accademia veneziana della Fama," in *Ethos e cultura: Studi in onore de Ezio Riondato,* 2 vols. (Padua: Antenore, 1991), 2:1133–1154.

22. *The Famous Historie of Fryer Bacon* (n.p., 189?), 63.

23. Robert Greene, *The Honorable Historie of frier Bacon and frier Bongay* (London: Printed for Edward White, 1594). On the play itself, see W. F. McNeir, "Traditional Elements in the Character of Greene's Friar Bacon," *Studies in Philology* 45 (1948): 172–179.

24. John Dee, "A very fruitfull praeface," in *The elements of geometrie of the most auncient philosopher Euclide of Megara,* trans. H. Billingsley (London: J. Daye, 1570), fols. a.4v–b.1r; cited in Van Helden, *Invention of the Telescope,* 29.

25. On the gift of the mirror, see Mordechai Feingold, *The Mathematicians' Apprenticeship: Science, Universities, and Society in England, 1560–1640* (Cambridge: Cambridge University Press, 1984), 157–158. The works featuring Dee's glass are both by Robert Armin, better known for his comedic roles in Shakespeare's plays than as a

writer: see *Two Maids of More-clack* (that is, Mortlake, where Dee lived) and *A Nest of Ninnies*. Armin is thought to have played the role of Sir Epicure Mammon, an old voluptuary whose use of a concave mirror for his amorous escapades derives from Seneca's *Quaestiones Naturales* 1:16, in Ben Jonson's *Alchemist* around 1610.

26. On Leonard and Thomas Digges, see the entries of Stephen Johnston in *Oxford Dictionary of National Biography*, ed. H. C. G. Matthew and Brian Harrison (Oxford: Oxford University Press, 2004), 16:169–170, 171–173.

27. See Harry E. Burton, "The *Optics* of Euclid," *Journal of the Optical Society of America* 35, no. 3 (1945): 357–372, on 360–361; and Euclid, *L'Optique et la Catoptrique*, trans., intro., and annotated by Paul ver Eecke (Paris: Albert Blanchard, 1959), xix. For a brief discussion of the important place of the two texts within the optical tradition, see A. Mark Smith, *Ptolemy and the Foundations of Ancient Mathematical Optics* (Philadelphia: American Philosophical Society, 1999), 11–19. On knowledge of Euclid's *Optics* before the early modern period, see Wilfred Theisen, "Euclid's *Optics* in the Medieval Curriculum," *Archives internationales d'histoire des sciences* 32 (1982): 159–176; on knowledge of that work in the early modern period, see Ver Ecke's commentary to Euclid's *L'Optique et la Catoptrique*, xxxvi–xliv.

28. Thomas Digges and Leonard Digges, *A Geometrical Practise, named Pantometria* (London: Henrie Bynneman, 1571), fols. G.i–ii.

29. I thank Al van Helden and Sven Dupré for pointing this out.

30. Thomas and Leonard Digges, *Stratioticos* (London: Henrie Bynneman, 1579), 190.

31. Thomas Birch, *The History of the Royal Society*, 4 vols. (London: A. Millar, 1756–1757), 4:157.

32. On Bourne, see G. L'E. Turner, "Bourne, William (c. 1535–1582)," in *Oxford Dictionary of National Biography*, 6:862–863; on the dat-

ing of his works, see Stephen Johnston, "A revised bibliography of William Bourne," www.mhs.ox.ac.uk/staff/saj/bourne/.

33. See "The 107: Devise," in William Bourne, *Inventions or Devises* (London: Printed by T. Woodcock, 1578), 92–94, and "The 122nd Devise" in the manuscript version, which may be consulted online through the University of Pennsylvania's Schoenberg Center for Electronic Text & Image, fols. 105r–106v. The annotation is on fol. 106v.

34. Bourne, *Inventions or Devises,* 96–97, cited in Van Helden, *Invention of the Telescope,* 30.

35. Bourne, *A Treatise on the Properties and Qualities of Glasses for Optical Purposes,* in *Rara mathematica,* ed. James O. Halliwell-Phillipps (London: J. W. Parker, 1839), 45–46; cited in Van Helden, *Invention of the Telescope,* 30–34.

36. Thomas Digges, whose son would become a well-known translator, referred to the use of the vernacular as protection in the closing sentences of the preface to his *Pantometria* and on the last page of the *Stratioticos,* p. 191. On the relatively low number of English speakers in this period, compared to speakers of French, German, Italian, and Spanish, see Peter Burke, *Languages and Communities in Early Modern Europe* (Cambridge: Cambridge University Press, 2004), 82.

37. Reginald Scot, *The Discovery of Witchcraft* (London: Andrew Clark, 1665), bk. 13, chap. 19, pp. 178–179. On the work itself, see Leland L. Estes, "Reginald Scot and his 'Discoverie of Witchcraft': Religion and Science in the Opposition to the European Witch Craze," *Church History* 52, no. 4 (1983): 444–456.

38. Scot, *The Discovery of Witchcraft,* 179.

39. On the project, see Eric Ash, "Expert Mediation and the Rebuilding of Dover Harbor," in *Power, Knowledge, and Expertise in Elizabethan England* (Baltimore: Johns Hopkins University Press,

2004), 55–86; and Stephen Johnston, "Dover Harbour: Constructing Design, Designing Construction," in *Making Mathematical Practice: Gentlemen, Practitioners and Artisans in Elizabethan England* (doctoral dissertation, Cambridge University, 1994), 218–226. For Scot's chronicle, see *Holinshed's Chronicles of England, Scotlande, and Irelande,* 6 vols. (London: J. Johnson, 1807–1808), 4:845–868; for Thomas Digges's report, see "A briefe Discourse on Dover Haven," *Archaeologia* 11 (1794): 212–254.

40. Edward Worsop, *A discouerie of sundrie errours* (London: Henrie Middleton for Gregorie Seton, 1582), fols. A2v, Gv, G2r, G3v, Kr.

41. Ibid., fol. Cr.

42. The association of *grimoire,* a magic manual, and *grammaire* was evident even to speakers of English, as the words *glamour* and *gramarye* suggest. Worsop's insistence on the analogy between the Euclidean elements of geometry and the rules of Latin grammar is presumably an effort to undo the original association. See ibid., fols. G2r–G3r.

43. Ibid., fol. Cv.

44. Even when more fully dissociated from the occult, the mathematical practitioner would be accused of merely executing a series of tricks without any sound theoretical knowledge of his discipline. See in this connection Katherine Hill, "'Juglers or Schollers?': Negotiating the Role of the Mathematical Practitioner," *British Journal for the History of Science* 31 (1998): 253–274.

45. Worsop, *Discouerie of sundrie errours,* fols. E3v, Cv.

46. Ibid., fol. Cv; Digges, *Pantometria,* fol. A2v.

47. For an excellent general introduction to della Porta, see Louise George Clubb, *Giambattista della Porta, Dramatist* (Princeton: Princeton University Press, 1965), 3–56.

48. "Lettera scritta al sopradetto M[aestro] Abramo Colorni con occasione del sonetto, et d'alcune annotationi antecedenti nella *Pi-*

azza," in Tomaso Garzoni, *La Piazza universale di tutte le professioni del mondo,* ed. Giovanni Battista Bronzino, Pina de Meo, and Luciano Carcereri, 2 vols. (Florence: Leo S. Olschki, 1996), 1:17–22, on 21. On Abramo Colorni generally, see C. Colombero, "Colorni, Abramo," in *Dizionario biografico degli italiani,* ed. Alberto M. Ghisalberti, 65 vols. to date (Rome: Istituto della Enciclopedia Italiana, 1960–), 27:466–468; and Iacopo Gaddi, *Adlocutiones* (Florence: Pietro Nesti, 1636), 167–169. For allusions to his catoptrical work in Colorni's letters from Prague in the late 1580s, see Giuseppe Jarè, "Abramo Colorno, ingegnere del secolo XVI," *Atti e memorie della deputazione provinciale ferrarese di storia patria* 3 (1891): 257–312, on 280, 282–283. The *Euthimetria,* licensed for publication by Duke Guglielmo Gonzaga of Mantua in 1580, exists only in manuscript form, no. 3046, in the Archducal Library at Wolfenbüttel; see A. Lewinski, "Alcuni manoscritti italiani nella Biblioteca Ducale di Wolfenbüttel," *Rivista Israelitica* 1 (1904): 236–237. To all appearances, Colorni did not deliver anything with telescopic capabilities; the device described on fols. 158r–v in the manuscript combines a large scaled instrument suitable for surveying variously with a magnetic compass, an odometer of Colorni's invention, and a plane mirror.

49. Luciano Chiappini, *Gli Estensi: Mille anni di storia* (Ferrara: Corbo Editore, 2001), 307–308.

50. Mario Gliozzi, "Relazioni scientifiche tra Paolo Sarpi e Giovan Battista della Porta," *Archives internationales d'histoire des sciences* 1 (1947): 395–433, on 419.

51. Paul Lawrence Rose, "Jacomo Contarini (1536–1595): A Venetian Patron and Collector of Mathematical Instruments and Books," *Physis* 18, no. 2 (1976): 117–130, on 124–125; and Michel Hochmann, "La collection de Giacomo Contarini," *Mélanges de l'Ecole française de Rome: Moyen Âge–Temps Modernes* 99, no. 1 (1987):

447–489, on 456; Giambattista della Porta, *Magia Naturalis,* bk. 7, proem.

52. Giambattista della Porta, *Magia naturalis libri viginti* (17, proem) (Naples: Orazio Salviani, 1589), 259.

53. F. Fiorentino, "Della vita e delle opere di G. B. della Porta," *Nuova antologia,* 2nd ser., 21 (1880): 251–294, on 267.

54. On the Pythagorean mirror in general, see Cornelius Agrippa, of Nettesheim, *Three Books of Occult Philosophy,* trans. J. F. (London: by R. W. for Gregory Moule, 1650), 16; Cœlius Rhodiginus, *Lectionum antiquarum libri triginta* (Cologne: Philippe Albert, 1620), lib. 9, cap. 23, col. 488; Natale Conti, *Mythologiae* (Padua: Pietro Paolo Tozzi, 1616), lib. 3, cap. 17, p. 133; Blaise de Vigenère, *Traicté des chiffres* (Paris: Abel l'Angelier, 1586), 16–17; Giovanni Battista della Porta, *Natural Magick,* lib. 17, cap. 17, p. 376; Isaac Beeckman, *Journal,* fol. 341v, in René Descartes, *Œuvres,* 10:347; August II [Gustavus Selenus], *Cryptomenytices et Cryptographiae Libri IX,* 1624, lib. 8, cap. 10, p. 425; Agostino Mascardi, *Saggi accademici dati in Roma nell'Accademia del Serenissimo Prencipe Cardinal di Savoia* (Venice: Bartolomeo Fontana, 1630), 161; Francis Godwin, *Nuncius Inanimatus,* and *The Man in the Moone,* 6, 47–48; John Wilkins, "Mercury; or the Secret and Swift Messenger," in *The Mathematical and Philosophical Works* (London: for J. Nicholson, 1708), cap. 19, pp. 77–78; Wilkins, *Discovery of a New World in the Moone,* in *The Mathematical and Philosophical Works,* 52; Gabriel Naudé, *The History of Magic,* trans. J. Davies (London: for J. Streater, 1657) 103, 107–108, 196; Jacques D'Autun, *L'Incredulité savante, et la credulité ignorante* (Lyon: Jean Molin, 1671), 992; Athanasius Kircher SJ, *Ars magna lucis et umbrae* (Amsterdam: Jansson-Wæsberg, 1671), 789–790; Gioseffo Petrucci, *Prodromo apologetico alli studi Chircheriani* (Amsterdam: Jansson-Wæsberg, 1677), 114.

55. Della Porta, *Magia naturalis libri viginti,* 17:11.270.

56. Johannes Kepler, *Dissertatio cum Nuncio Sidereo,* in Galileo Galilei, *Opere,* ed. Antonio Favaro, 20 vols. (Florence: Giunti Barbèra, 1968), 3(1):109, 13:212–213.

57. Clubb, *Giambattista della Porta,* 234.

58. Plautus, *Miles Gloriosus* 2.3.

59. Giambattista della Porta, *La Chiappinaria,* 3.1.13–15; and 3.3.26–32 in Giambattista della Porta, *Teatro,* ed. Raffaele Sirri, 4 vols. (Naples: Edizioni Scientifiche Italiane, 2003), 4:44 and 48.

60. Bonaventura Cavalieri, *Lo specchio ustorio* (Bologna: Ferroni, 1632), 126. Cavalieri goes on to add to the "Ptolemaic" combination a flat mirror, aligned with the ocular and at an angle to the concave objective. The addition of this third element suggests (and is) an updated improvement over the configuration associated with Ptolemy. See Piero Ariotti, "Bonaventura Cavalieri, Marin Mersenne, and the Reflecting Telescope," *Isis* 66 (1975): 303–321, on 314–316.

61. Giambattista della Porta, *De Refractione optices parte: Libri novem,* 1:11, 1:12, 2:16, 5:16, 8:3, 8:19 (Naples: Orazio Salviani, 1593), 20, 24, 52, 126–127, 176, 188. The story about the sharp-sighted man in Sicily would be appropriated by Francesco Stelluti, the dedicatee of *Chiappinaria* and an early member of the Academy of the Lynxes; see C. H. Lüthy, "Atomism, Lynceus, and the Fate of Seventeenth-Century Microscopy," *Early Science and Medicine* 1 (1996): 1–27, on 7.

62. Giuseppe Gabrieli, "Giovan Battista della Porta Linceo," *Giornale critico della filosofia italiana* 8 (1927): 360–397, on 362, 369–371.

63. Edmund Spenser, *The Faerie Queene,* 3:2:19.

64. Ibid., 3:2:20. The reinscription of such devices into early modern epics and romances was relatively common in this period. Around 1650, for example, Sir William Davenant described the "Optick Tubes" through which the subjects of the wise Astragon observe

the moon as an invention then "nine hasty Centuries" old; see Sir William Davenant, *Gondibert,* bk. 2, canto 5, 16–17. John Milton alluded several times to Galileo's "optic tube" in *Paradise Lost,* and to the more mysterious "airy microscope" in *Paradise Regained;* see *Paradise Lost,* 3:590, and *Paradise Regained,* 4:40–42, 56–57. The telescope appears in numerous Italian romances from the 1620s until about 1650. Not surprisingly, these devices retain something of the imperial instrument that preceded them; in Giulio Strozzi's *Venetia Aedificata* of 1624, for instance, Merlin is the inventor of the telescope. See Antonio Belloni, *Gli epigoni della Gerusalemme Liberata* (Padua: A. Draghi, 1893), 207, 312.

3. Obscure Procedures and Odd Opponents

1. See Mario Gliozzi, "Relazioni scientifiche tra Fra Paolo Sarpi e Giovan Battista Porta," *Archives internationales d'histoire des sciences* 1 (1947): 395–433; and Luisa Cozzi, "La formazione culturale e religiosa e la maturazione filosofica e politico-giuridica nei *Pensieri* di Paolo Sarpi," in Paolo Sarpi, *Pensieri naturali, metafisici e matematici,* ed. and annotated by Luisa Cozzi and Libero Sosio (Milan: Riccardo Ricciardi, 1996), xxv–lxxxviii.

2. Sarpi, *Pensieri,* 82–84, 190–191, 207, 272, 372, 436.

3. Ibid., 84. All translations mine, unless otherwise indicated.

4. On the observation well, see François Arago, *Astronomie populaire* (Paris: Gide, 1857), 202–203; Robert Eisler, "The Polar Sighting-Tube," *Archives internationales d' histoire des sciences* 2 (1948): 312–332, on 313–316; and Aydin Sayili, "The 'Observation Well,'" *Dil ve Tarih-Coğrafya Fakültesi Dergisi* (Ankara University) 11 (1953): 149–155. Thomas Tomkis, the first playwright to put the Dutch telescope on the English stage with his *Albumazar,* ca. 1615, compared the clenched fist to a perspective glass in a play of 1607; see

Tomkis, *Lingua: Or the Combat of the Tongue, and the Five Senses for Superiority* 1:7 (New York: AMS Press, 1970), fol. B4.

5. Sarpi, *Pensieri,* 18, 93, 233, 289.

6. Aristotle, *On the Generation of Animals,* 5:1 (780b–781a), trans. A. I. Peck (Cambridge, MA: Harvard University Press, 1953), 503–505.

7. See Galileo Galilei, *Opere,* ed. Antonio Favaro, 20 vols. (Florence: Giunti Barbèra, 1968), 10:285–286, 341–342; and Albert van Helden, *The Invention of the Telescope* (Philadelphia: American Philosophical Society, 1977), 35, 45–46.

8. Giambattista della Porta, *Dei miracoli et maravigliosi effetti* (Venice: L. Avanzi, 1560), 140–142.

9. Sarpi, *Pensieri,* 436; for a transcription with slight but crucial differences, see Gliozzi, "Relazioni," 415. The Venetian pace equaled 1.73 meters; see Sarpi, *Pensieri,* 511, n. 673.

10. See, for instance, during the debate over the sunspots, Georg Stengel SJ to Karl Stengel, in Fidel Rädle, "Die Briefe des Jesuiten Georg Stengel (1584–1651) an seinen Bruder Karl (1581–1663)," in *Res Publica Litteraria: Die Institutionen der Gelehrsamkeit in der frühen Neuzeit,* ed. Sebastian Neumeister and Conrad Wiedemann (Wiesbaden: Otto Harrassowitz, 1987), 2:525–534, on 530–531.

11. "Il Manoscritto dell'Iride e del Calore," in Sarpi, *Pensieri,* 529–546, prefaced by the observations of Sosio, 519–528.

12. Cozzi, "La formazione culturale," xxxiii–xxxiv.

13. Sven Dupré, "Ausonio's Mirrors and Galileo's Lenses: The Telescope and Sixteenth-Century Practical Optical Knowledge," *Galilaeana* 2 (2005): 145–180, on 167–168.

14. For an overview of such attitudes, see the helpful discussion in Dupré, "Ausonio's Mirrors," 146–147.

15. See Luigi Firpo, "Appunti Campanelliani: Storia di un furto," *Giornale critico della filosofia italiana* 10 (1956): 541–549, on 545–546; and John M. Headley, *Tommaso Campanella and the Transfor-*

mation of the World (Princeton: Princeton University Press, 1997), 27–28.

16. Dupré, "Ausonio's Mirrors," 160–170.

17. Galileo, *Opere,* 10:54, 55–60, 62.

18. Ibid., 19:117–125; and Mario Biagioli, *Galileo's Instruments of Credit: Telescopes, Images, Secrecy* (Chicago: University of Chicago Press, 2006), 7–12.

19. Galileo, *Opere,* 19:149–150.

20. Vincenzio Viviani, *Racconto istorico,* in ibid., 19:597–632, on 606.

21. On Seget, see Antonio Favaro, "Dall' 'Album Amicorum' di Tommaso Seggett," *Atti e memorie della Reale Accademia di scienze, lettere ed arti in Padova,* n.s., 6 (189): 58–62; Favaro, "Amici e corrispondenti di Galileo Galilei: Tommaso Segeth," *Atti del Reale Istituto Veneto di scienze, lettere ed arti* 70, no. 2 (1910–1911): 617–654; Otakar Odlozilik, "Thomas Seget: A Scottish Friend of Szymon Szymonowicz," *Polish Review* 11, no. 1 (1966): 3–39; and Ö. Szabolcs Barlay, "Thomas Seget's (from Edinborough) Middle European Connections in Reflection of Cod. Vat. Lat. 9385," *Magyar Könyvszemle* 97, no. 3 (1981): 204–220. On the confusing and ultimately disastrous fate of Pinelli's library, see Marcella Grendler, "A Greek Collection in Padua: The Library of Gian Vincenzo Pinelli (1535–1601)," *Renaissance Quarterly* 33, no. 3 (1980): 386–416, particularly 388–391.

22. Favaro, "Dall' 'Album Amicorum,'" 60; Odlozilik, "Thomas Seget," 10 and n. 29; and Barlay, "Thomas Seget's Middle European Connections," 206–209.

23. Galileo, *Opere,* 19:203–204; and Barlay, "Thomas Seget's Middle European Connections," 209. Galileo used the figure of the parabola, with somewhat different implications, in at least one other *album amicorum;* see his *Opere,* 19:204. For remarks on the connection of the rhetorical figures of hyperbole, parable, and ellipsis with

their geometrical analogues, see Galileo, *Opere,* 4:467, 698. In a letter of 1580 to his patron, della Porta likewise exploited the pun,
claiming that the failure of his parabolic mirror would make a parable of him. See Fiorentino, "Della vita e delle opera di G. B. della
Porta," *Nuova antologia,* 2nd ser., 21 (1880): 251–294. For Galileo's
interest in parabolic trajectories, see Jürgen Renn, Peter Damerow,
and Simone Rieger, with an appendix by Domenico Giulini,
"Hunting the White Elephant: When and How Did Galileo Discover the Law of Fall?" in *Galileo in Context,* ed. Jürgen Renn
(Cambridge: Cambridge University Press, 2001), 29–149, esp. 54–57.

24. On Pigafetta, see Wolfram Prinz, "Informazione di Filippo
Pigafetta al Serenissimo di Toscana per una stanza da piantare lo
studio di architettura militare," *Atti del convegno internazionale di
studi* (Florence: Leo S. Olschki, 1983), 1:343–353; and Prinz, "Dal
modello al dipinto: Macchine da guerra di Archimede alla fine del
Cinquecento," in *Architettura militare nell'Europa del XVI secolo,*
ed. Carlo Cresti, Amelio Fara, and Daniela Lamberini (Siena:
Edizioni Periccioli, 1988), 409–416. On Marino Ghetaldi, see
Florio Banfi, "Marino Ghetaldi da Ragusa e Tommasso Seghet da
Edimburgo," *Archivio storico per la Dalmazia* 26 (1939): 322–345;
and Ernest Stipanić, "La vita e il lavoro di Marin Getaldić
(Marinus Ghetaldus)," *Dijalektika* 19 (1984): 5–21.

25. Galileo, *Opere,* 10:94–95.

26. See Fabio Mutinelli, *Storia arcana ed aneddotica d'Italia raccontata
dai Veneti ambasciatori,* 4 vols. (Venice: Pietro Naratovich, 1858),
3:254–255; Carlo Promis, *Gl'ingegneri militari che operarono o
scrissero in Piemonte dall'anno MCCC all'anno MDCL* (Bologna:
Forni Editore, 1973), 52–54; S. Meschini, "Gromo, Giacomo Antonio," in *Dizionario biografico degli italiani,* ed. Alberto M.
Ghisalberti, 65 vols. to date (Rome: Istituto della Enciclopedia
Italiana, 1960–), 59:766–768; Luciana Bona Quaglia and Sergio

Tira, "*Gromida:* Alchimia e versificazione latina in un manoscritto torinese del primo Seicento," *Studi Piemontesi* 23, no. 1 (1994): 23–58; and Franco Tomasi, "La malagevolezza delle stampe: Per una storia dell'edizione Discepolo del 'Mondo Creato,'" *Studi Tassiani* 42 (1994): 43–78; PRO, *Calendar of State Papers, Foreign Series* (London: HMSO, 1861–1950), 10:62–63, 285–286. For an informative overview of the milieu in which Gromo flourished, see William Eamon, *Science and the Secrets of Nature: Books of Secrets in Medieval and Early Modern Culture* (Princeton: Princeton University Press, 1994).

27. Angelo Ingegneri, *Contra l'alchimia, e gli alchimisti: Palinodia dell'Argonautica* (Naples: Gio. Giacomo Carlino, 1606), 59. The comparison of Galileo's name and surname to Galilee was commonplace.

28. See Raphael Thorius, *Hymnus tabaci* (London: John Waterson, 1626), 5–6; Tomaso Garzoni, *La Piazza universale di tutte le professioni del mondo,* ed. Giovanni Battista Bronzino, Pina de Meo, and Luciano Carcereri, 2 vols. (Florence: Leo S. Olschki, 1996), 1094; and Athanasius Kircher SJ, *Ars magna lucis et umbrae* (Amsterdam: Janson, 1671) 424. More generally on interpretations of the Promethean story, see Olga Raggio, "The Myth of Prometheus: Its Survival and Metamorphoses up to the Eighteenth Century," *Journal of the Warburg and Courtauld Institutes* 21, nos. 1–2 (1958): 44–62.

29. Filippo Pigafetta, *Le mechaniche dell'Ill. Sig. Guido Ubaldo de' Marchesi del Monte* (Venice: F. di Franceschi, 1581), blind folio 2v.

30. Garzoni, *Piazza universale,* 931; and Bartolomeo Crescentio, *Nautica mediterranea* (Rome: Bartolomeo Bonfadino, 1602), 46. Galileo, among others, had used the Archimedean screw to design a water-lifting device; see Stillman Drake, *Galileo at Work: His Scientific Biography* (Chicago: University of Chicago Press, 1978), 35, 61.

31. Ingegneri, *Contra l'alchimia*, 58.
32. See A. Siekiera, "Ingegneri, Angelo," in *Dizionario biografico degli italiani*, 62:358–361; Torquato Tasso, *Lettere*, ed. Cesare Guasti, 4 vols. (Florence: Le Monnier, 1854), 4:164–165, 171–172, 209–212; and Louise George Clubb, *Giambattista della Porta, Dramatist* (Princeton: Princeton University Press, 1965), 7, 9, 126–128. Ingegneri and Pigafetta were associated with rival editions of one of Tasso's works, and they were both involved in the famous production of *Edipo Tiranno* in Vicenza in 1585.
33. Compare Galileo, *Il Compasso Geometrico e Militare*, in *Opere*, 2:335–424, on 403–404, 412–424; and Bona Quaglia and Tira, *"Gromida,"* 24–25, 46–49, 54–55.
34. Mario Pozzi, "Appunti su Filippo Pigafetta," in *Miscellanea di studi in onore di Claudio Varese*, ed. Giorgio Cerboni Baiardi (Rome: Vecchiarelli, 2001), 635–656, on 644, 650, 652–653.
35. Ingegneri, *Contra l'alchimia*, 63–64; Promis, *Gl'ingegneri*, 54; and Galileo, *Difesa contro alle calunnie ed imposture di Baldessar Capra*, in *Opere*, 2:537.
36. Galileo, *Opere*, 10:106. The National Edition mistakenly records the name here as "Grosso."
37. See Kate van Orden, *Music, Discipline, and Arms in Early Modern France* (Chicago: University of Chicago Press, 2005), 57–62; and Francisco de Quevedo, *La Vida del Buscon*, ed. Fernando Cabo Aseguinolaza (Barcelona, Crítica, 1993), 109.
38. On the legal, financial, and social aspects of the whole episode—from Capra's examination of the compass in the spring of 1605, to Galileo's decision to issue a small printed version of his instructions on the use of the instrument in 1606, to Capra's publication, the trial, and Galileo's *Defense* in the spring and summer of 1607—see Mario Biagioli, "Galileo v. Capra: Intellectual Property between Paper and Brass," *History of Science* (forthcoming). I am grateful to Professor Biagioli for allowing me to consult this essay.

39. Stillman Drake, "Tartaglia's *Squadra* and Galileo's *Compasso,*" *Annali dell'Istituto e Museo di Storia della Scienza* 2 (1977): 35–54. On the development of the proportional compass, see Filippo Camerota, *Il Compasso di Fabrizio Mordente: Per la storia del compasso di proporzione* (Florence: Leo S. Olschki, 2000).

40. Baldessar Capra, *Usus et Fabrica Circini,* in Galileo, *Opere,* 2:425–510, on 490; and Galileo, *Difesa,* in *Opere,* 2:423–424.

41. Galileo, *Difesa,* in *Opere,* 2:519, 545, 594. On those who cannot be the "ancient adversary"—Mayr, Capra, and Magini—see Stillman Drake, "Was Simon Mayr Galileo's 'Ancient Adversary' in 1607?" *Isis* 67 (1976): 456–462.

42. Ingegneri, *Contra l'alchimia,* 10, 16; and Galileo, *Difesa,* in *Opere,* 2:518, 587, 590.

43. Galileo, *Difesa,* in *Opere,* 2:519, 545, 559, 564, 585, 594.

44. Ibid., 2:581–582.

45. Capra, *Usus,* in Galileo, *Opere,* 2:492; and Galileo, *Difesa,* in *Opere,* 2:584–585.

46. Galileo, *Difesa,* in *Opere,* 2:534–535, 539, and 19:224–226.

47. Ibid., 2:582, 583, 584, 586, 587, 588, 589–590, 591, 592.

48. On this and Capra's other plagiarisms, see Michele Camerota, *Galileo Galilei e la cultura scientifica nell'età della controriforma* (Rome: Salerno Editrice, 2004), 122–130.

49. See Eileen Reeves, *Painting the Heavens: Art and Science in the Age of Galileo* (Princeton: Princeton University Press, 1997), 91–137; and Sven Dupré, "Galileo's Telescope and Celestial Light," *Journal for the History of Astronomy* 34 (2003): 369–399.

50. Galileo, *Difesa,* in *Opere,* 2:519; for a typical version of the tale of the basilisk, see story 22 of the *Gesta Romanorum* (London: T. Norris, 1722), 77.

51. Galileo, *Difesa,* in *Opere,* 2:521 (emphasis mine).

52. Galileo, *Opere,* 10:346, 357, 381.

53. Ibid., 12:195–196; on Danti's instruments, see Guglielmo Righini,

"Il grande astrolabio del Museo di Storia della Scienza di Firenze," *Annali dell'Istituto e Museo di Storia della Scienza* 2, no. 2 (1977): 45–66; Maria Luisa Righini Bonelli and Thomas Settle, "Egnatio Danti's Great Astronomical Quadrant," *Annali dell'Istituto e Museo di Storia della Scienza* 4, no. 2 (1979): 3–13; and J. L. Heilbron, *The Sun in the Church: Cathedrals as Observatories* (Cambridge, MA: Harvard University Press, 1999), 47–81.

54. *Vita scritta da Niccolò Gherardini,* in Galileo, *Opere,* 19:633–646, on 637–638.

4. *The Dutch Telescope and the French Mirror*

1. On the embassy and on early relations between the Dutch and the Siamese, see Dirk Van der Cruysse, *Louis XIV et le Siam* (Paris: Fayard, 1991), 53–69. For a transcription of the pamphlet and some discussion of the embassy, see Paul Pelliot, "Les relations du Siam et de la Hollande en 1608," *T'oung Pao* 32 (1936): 223–229; for the aftermath of the encounter with Maurice of Nassau, see J. J. L. Duyvendak, "The First Siamese Embassy to Holland," *T'oung Pao* 32 (1936): 285–292.

2. For manuscript versions, see Vatican Library, Fondo Urbinate Latino 1080, Avvisi manoscritti, fols. 721r, 725v. For a reproduction of the entire newsletter and discussion of the brief article on the telescope, see Stillman Drake, *The Unsung Journalist and the Origin of the Telescope* (Los Angeles: Zeitlin and Ver Brugge, 1976). For a valuable discussion of the vexed issue of the Dutch telescope's invention, see Albert Van Helden, *The Invention of the Telescope* (Philadelphia: American Philosophical Society, 1977).

3. Paolo Sarpi, *Lettere ai protestanti,* ed. Manlio Duilio Busnelli, 2 vols. (Bari: Giuseppe Laterza & Figli, 1931), 1:58. All translations are mine, unless otherwise indicated.

4. Manlio Busnelli, "Un carteggio inedito di Fra Paolo Sarpi con

l'Ugonotto Francesco Castrino," in *Études sur Fra Paolo Sarpi* (Geneva: Slatkine, 1986), 33–169, on 75.

5. Subsequent to its emergence in Latin and the vernacular in 1608, the *Discoverie* was cited in the *Mercurius Gallobelgicus* in 1609, incorporated into anti-Jesuit sermons and writings by Lutheran preachers in Augsburg in 1609 and 1610, presented in an anthology of anti-Jesuit pamphlets in 1610, refuted in the *Annual Letter* of the Society of Jesus in 1610 and by Jacob Gretser SJ in 1609, 1610, and 1612, and reissued from Basel in 1627.

6. Sarpi, *Lettere ai protestanti,* 1:34–35, 47.

7. [Johannes Cambilhom], *A Discoverie of the most secret and subtile practices of the* IESVITES, *Translated out of French* (London: Robert Bolton, 1610), fol. B2r.

8. *Le Supplément du Catholicon, ou Nouvelles des régions de la lune,* in *Satyre Menippée de la vertu du catholicon d'Espagne,* ed. Charles Nodier, 2 vols. (Paris: N. Delangle, 1824), 2:264, 307.

9. Gabriel Naudé, *Apologie pour tous les grands personnages qui ont esté faussement soupçonnez de magie* (The Hague: Adrian Vlac, 1653), 493–494. Naudé appears to suggest an amusing connection between ceremonies honoring, or aggrandizing, Ignatius of Loyola and Francis Xavier and those feats involving magnification and the apparent metamorphosis of men into swine and asses. The work in catoptrics at Pont-à-Mousson was associated with Jean Leurechon, whose presence is attested at the university in that period, and under whose guidance the *Récréations mathématiques,* devoted in part to catoptrics, would be published in 1624. A phrase not unlike that cited by Naudé appears in a discussion of "Des autres miroirs de plaisir" and is qualified by the observation that "this application is the work of angelic, rather than human, subtlety." See Jacques Ozanam, *Récréations mathématiques,* 6th ed. (Rouen: C. Mallasis, 1669), 243. On the complicated problem of the authorship and var-

ious editions of this work, see Trevor H. Hall, *Mathematicall Recreations: An Exercise in Seventeenth-Century Bibliography* (Leeds: University of Leeds Press, 1969). On Leurechon's courses, see G. Gavet, ed., *Diarium universitatis mussipontanae (1572–1764)* (Paris: Berger-Levrault, 1911), cols. 120–121, 128, 144, 146, 151, 161, 170, 175, 189. The 1622 ceremonies and publication mentioned by Naudé are described in the *Diarium* under the year 1623; see *Diarium,* col. 153.

10. On Grienberger's and Guevara's work on elliptical mirrors, see the assessment of Bonaventura Cavalieri in *Lo specchi ustorio* (Bologna: Clemente Ferroni, 1632), 78–84.

11. *Recueil des letters missives de Henri IV,* ed. M. Berger de Xivrey (Paris: Imprimerie Impériale, 1853), 6:542–544.

12. Pierre de l'Estoile, *Mémoires-Journaux 1574–1611,* ed. G. Brunet et al., 12 vols. (Paris: Librairie des Bibliophiles, 1881), 8:191.

13. *Le Passe-partout des Pères Iésuites* (n.p., 1607), 160.

14. *Epistres françoises des personnages illustres & doctes, à Monsr. Ioseph Iuste de la Scala,* ed. Iaques de Reves (Harderwyck: Thomas Henry, 1624), 422, 432.

15. Jean Marie Prat, *Recherches historiques et critiques sur la Compagnie de Jésus en France au temps de P. Coton, 1564–1626,* 5 vols. (Lyon: Briday, 1876–1878), 2:437–438, 5:237.

16. Cambilhom, *A Discoverie,* fol. B2.

17. [Anonymous], *Le Remerciment des Beurrieres de Paris, au Sieur de Courbouzon de Mongommery* (Sedan: Guion de la Plume, 1610), 20.

18. L'Estoile, *Mémoires-Journaux,* 9:197–198.

19. Ibid., 10:204–205.

20. Sarpi, *Lettere ai protestanti,* 1:34–35; Prat, *Recherches historiques et critiques,* 3:131–133; Boris Ulianich, "Saggio Introduttivo," in Sarpi, *Lettere ai gallicani,* xli–xliv.

21. L'Estoile, *Mémoires-Journaux,* 10:30, 43.

22. Ibid., 10:53–54.

23. *Satyre Menipée,* 1:vi; Sarpi, *Lettere ai gallicani,* ed. Boris Ulianich (Wiesbaden: Franz Steiner, 1961), lxxxviii, 84, 268 n. 83.

24. Luigi Lazzerini, "Officina Sarpiana: Scritture del Sarpi in materia di Gesuiti," *Rivista di Storia della Chiesa in Italia* 58, no. 1 (2004): 29–80.

25. *Carteggio inedito di Ticone Brahe, Giovanni Keplero e di altri celebri astronomi e matematici dei secoli XVI e XVII con Giovanni Antonio Magini,* ed. Antonio Favaro (Bologna: Nicola Zanichelli, 1886), 448–449.

26. Nicolas Vauquelin des Yveteaux, *Oeuvres Complètes,* ed. Georges Mongrédien, 2 vols. (Paris: Auguste Picard, 1921), 2:164.

27. Archimedes, *Opera quae extant novis demonstrationibus comment-ariisque illustrata,* ed. David Rivault, Sieur de Fleury (Paris: Claude Morellus, 1615), 546–547. See D. L. Simms, "Galen on Archimedes: Burning Mirror or Burning Pitch?" *Technology and Culture* 32 (1991): 91–96. On Rivault's career as a humanist, see James J. Supple, "The Failure of Humanist Education: David de Fleurance-Rivault, Antoine Mathé de Laval, and Nicolas Faret," in *Humanism in Crisis,* ed. Philippe Desan (Ann Arbor: University of Michigan Press, 1991), 35–53, on 37–40.

28. On Badovere, see Antonio Favaro, "Amici e corrispondenti di Galileo Galilei: Giacomo Badouère," *Atti del Reale Istituto Veneto di scienze, lettere ed arti* 65, no. 2 (1905): 193–201; B. Ulianich, "Badoer, Giacomo," in *Dizionario biografico degli italiani,* ed. Alberto M. Ghisalberti, 65 vols. to date (Rome: Istituto della Enciclopedia Italiana, 1960), 5:114–116; and Franco Musarra, "Giacomo Badovere e il problema dei 'Libertini,'" *Ateneo Veneto,* n.s., 11, nos. 1–2 (1973): 121–137.

29. On this friendship, see Musarra, "Giacomo Badovere"; and *Briefwisseling van Pieter Corneliszoon Hooft,* ed. H. W. van Tricht, 3

vols. (Culemborg: Tjeenk Willink / Noorduijn, 1976), 1:72–75, 76–78, 84–88, 91–98, 104–108, 126–127.

30. Musarra, "Giacomo Badovere," 122. Passing mention is made of Badovere's role in *Recueil des letters missives de Henri IV,* 6:88, 111, and 9:148, 241, 242.

31. See Favaro, "Amici e corrispondenti," 195–196; and Jacques Pannier, "Pasteurs et autres protestants convertis et pensionnés par le clergé de 1603 à 1617," *Bulletin de la société de l'histoire du protestantisme français,* ser. 5, vol. 6 (1907): 233–263; "Jacob Badouere" is mentioned on pp. 246 and 251, in regard to *pensions* collected in 1610–1612.

32. Sarpi, *Lettere ai gallicani,* 163.

33. Galileo Galilei, *Difesa contro alle calunnie ed imposture di Baldessar Capra,* in *Opere,* ed. Antonio Favaro, 20 vols. (Florence: Giunti Barbèra, 1965), 2:534.

34. See the vicious lines written around January 1610 by Sarpi's correspondent Francesco Castrino and recorded in part by Pierre de l'Estoile in his *Mémoires-Journaux,* 10:120, 125–126. On the accusation of spying, see the conjecture of Antonio Favaro, "Amici e corrispondenti di Galileo Galilei: Martino Hasdale," *Atti del Reale Istituto Veneto di scienze, lettere ed art* 55, no. 2 (1905): 202–208, on 206; and the assertion of Boris Ulianich, "Badoer, Giacomo," 114–116, based on the archival work of Pietro Savio, "Per l'epistolario di Paolo Sarpi," *Aevum* 10 (1936): 3–104, and 11 (1937): 13–74, 275–322.

35. Pietro Savio, "Per l'epistolario di Paolo Sarpi," *Aevum* 10 (1936): 3–104, and 11 (1937): 13–74, 275–322, in 10:29, 33.

36. Sarpi, *Lettere ai protestanti,* 2:49; and l'Estoile, *Mémoires-Journaux,* 10:126.

37. [Marie de Gournay], *Adieu de l'âme du Roy de France et de Navarre*

Henry le Grand, à la Royne, in *Textes relatifs à la calomnie,* ed. Constant Venesoen (Tübingen: Gunter Narr, 1998), 44. On Sarpi's growing suspicion that Badovere, not Mademoiselle de Gournay, was the author of the work, compare *Lettere ai protestanti,* 1:145 and 2:110.

38. Van Tricht, *De Briefwisseling van Pieter Corneliszoon Hooft,* 1:91, 106.

39. François d'Aerssen's father and Pierre Jeannin had news of the Dutch telescope within days of its invention in The Hague. See Van Helden, *Invention of the Telescope,* 25–26, 39–40, 43; and Van Tricht, *De Briefwisseling van Pieter Corneliszoon Hooft,* 1:96, 104, 106.

40. Galileo, *Opere,* 11:321; and Sarpi, *Lettere ai gallicani,* 203, 204.

41. The letters themselves, the exchanges of Borghese and Ubaldini, and some details about the manner in which the missives were obtained are in Savio, "Per l'epistolario di Paolo Sarpi," *Aevum* 10 (1936): 3–104, and 11 (1937): 13–74, 275–322. On the case of Francesco Castrino's correspondence with Sarpi, see Busnelli, "Un carteggio inedito," on 43–46. On the limited utility of these purloined letters to the papal nuncio, see Jonathan Walker, with Filippo de Vivo and James Shaw, "A Dialogue on Spying in 17th-century Venice," *Rethinking History* 10, no. 3 (2006): 323–344, on 331–332, and 341 n. 3.

42. Sarpi, *Lettere ai protestanti,* 1:72, 79.

43. In his letter of December 1607 to Hooft, Badovere stated that he was beginning to recover from fifteen months of ill health. Van Tricht, *De Briefwisseling van Pieter Corneliszoon Hooft,* 1:94.

44. Sarpi, *Lettere ai gallicani,* 179.

45. Ibid.

46. Seneca, *Epistulae morales* 68.3. The paraphrase is more ample and accurate in Sarpi's "Pensieri medico-morali": "Do not make what

you do in your moments of leisure public, but in order to escape envy, disguise it as either sickness or ineptitude, and above all not as philosophy, but rather as a rest for the mind, as long as this does not detract from your other affairs." See Paolo Sarpi, "Pensieri medico-morali," in *Pensieri naturali, metafisici e matematici,* ed. Luisa Cozzi and Liberio Sosio (Milan: Riccardo Ricciardi, 1996), 617.

47. Sarpi, *Lettere ai gallicani,* 180.

48. On the observation well, see François Arago, *Astronomie populaire* (Paris: Gide, 1857), 202–203; Robert Eisler, "The Polar Sighting-Tube," *Archives internationales d' histoire des sciences* 2 (1948): 312–332, on 313–316; and Aydin Sayili, "The 'Observation Well,'" *Dil ve Tarih-Coğrafya Fakültesi Dergisi* (Ankara University) 11 (1953): 149–155.

49. Sven Dupré, "Galileo's Telescope and Celestial Light," *Journal for the History of Astronomy,* 34 (2003): 369–399.

50. Galileo, *Opere,* 10:380–381.

51. Savio, "Per l'Epistolario di Paolo Sarpi," *Aevum* 11 (1937): 76. Curiously, it was in another letter, one written to Foscarini in late April 1609, that Sarpi stated he wanted to undertake the trip to France in order to carry from Venice arguments that would appear "as clear as if they were in a mirror"; see Savio, "Epistolario," 10:65.

52. Galileo, *Opere,* 10:225, 226, 229–230, 232–234, 236, 250–251.

53. Ibid., 10:230, 232–233.

54. Stillman Drake, *Galileo at Work: His Scientific Biography* (Chicago: University of Chicago Press, 1978), 134–135.

55. Galileo, *Opere,* 10:203, 204, 262; Gaetano Cozzi, *Paolo Sarpi tra Venezia e l'Europa* (Turin: Einaudi, 1979), 175–177; and Savio, "Epistolario," 11:31–32.

56. Galileo, *Opere,* 10:261.

57. Sarpi, *Lettere ai gallicani,* 229.

58. Galileo, *Opere,* 10:230.

59. Galileo Galilei, *Sidereus Nuncius,* trans., with introduction, conclusion, and notes, by Albert van Helden (Chicago: University of Chicago Press, 1989), 37.

60. Galileo, *Opere,* 10:250, 251; Galileo, *Sidereus Nuncius,* 6–9.

61. See Ewan Whitaker, "Galileo's Lunar Observations and the Dating of the Composition of 'Sidereus Nuncius,'" *Journal for the History of Astronomy* 9 (1978): 155–169; and Owen Gingerich and Albert van Helden, "From *Occhiale* to Printed Page: The Making of Galileo's *Sidereus Nuncius,*" *Journal for the History of Astronomy* 34 (2003): 251–267.

62. On Galileo's notions of the moon's rocky surface, ashen light, and mantle of vapor in the pre-telescopic period, see Eileen Reeves, *Painting the Heavens: Art and Science in the Age of Galileo* (Princeton: Princeton University Press, 1997), 3–137.

63. Galileo, *Opere,* 11:259.

64. See Van Helden's comments in *Sidereus Nuncius,* 59–61 nn. 63–65.

65. Colin A. Ronan, "The Origins of the Reflecting Telescope," *Journal of the British Astronomical Association* 1010 (1991): 335–342, on 341–342.

66. The references are to Propertius, *Elegies,* 3.2.19, *nam neque Pyramidum sumptus ad sidera ducti,* and to Horace, *Odes* 3.30.1–2, *Exegi monumentum . . . regali situ Pyramidum altius.* For an excellent Latin and Italian edition of *Starry Messenger,* complete with references to the several classical allusions in this part of the treatise, see Galileo Galilei, *Sidereus Nuncius,* ed. Andrea Battistini, trans. Maria Timpanaro Cardini (Venice: Marsilio, 1993).

67. I have followed the translation of Van Helden, *Sidereus Nuncius,* 29–31.

68. Projection within the camera obscura was routinely associated with printing and engraving, for the technique allowed one to make

fleeting exterior phenomena more or less permanent. This sense is implied by the verb *pulsetur,* "to be struck," though in his later works Galileo would use the terms *incidere* or *stampare,* more specific to printing, when discussing projection.

5. The Afterlife of a Legend

1. See Giovanni Baffetti, "Il 'Sidereus Nuncius' a Bologna," *Intersezioni* 11, no. 3 (1991): 477–500; and Isabelle Pantin, *Discussion avec le messager céleste* (Paris: Belles Lettres, 1993), xxviii–lxxviii. On Magini's astrology and astronomy, see Gian Luigi Betti, "Il copernicanesimo nello Studio di Bologna," and Enrico Peruzii, "Critica e rielaborazione del sistema copernicano in Giovanni Antonio Magini," both in *La diffusione del copernicanesimo in Italia (1543–1610),* ed. Massimo Bucciantini and Maurizio Torrini (Florence: Leo S. Olschki, 1997), 67–81, 83–98.

2. Benjamin Martin, *Biographia philosophica* (London: W. Owen, 1764), 511 (*rectè* 211)–212.

3. Giovanni Antonio Magini, *De dimetiendi ratione per quadrantem, & geometricum quadratum, libri quinque* (lib. 2, props. 30, 32, 33) (Venice: Roberto Meietti, 1592), fols. 87r, 89r, 92v). See also Fabrizio Bònoli and Marina Zuccoli, "On Two Sixteenth-Century Instruments by Giovanni Antonio Magini (1555–1617)," *Nuncius* 14 (1999): 201–212. For Capra's appropriation of Magini's work, see Galileo, *Opere,* ed. Antonio Favaro, 20 vols. (Florence: Giunti Barbèra, 1977), 10:174, 204, and 2:495–510, 593–598.

4. On his edition of Ausonio's work, see Sven Dupré, "Mathematical Instruments and the 'Theory of the Concave Spherical Mirror': Galileo's Optics beyond Art and Science," *Nuncius* 15 (2000): 551–588, on 563–572; Giovanni Antonio Magini, *Briefve instruction,* trans. Jean-Jacques Boyssier (Paris: n.p., 1620), 7–8.

5. Robert Burton, *The Anatomy of Melancholy,* intro. William H. Gass (New York: New York Review of Books, 2001), 2:2:4, 96. On Claude Mydorge, see Jean-François Gauvin, "Artisans, Machines, and Descartes's *Organon,*" *History of Science* 44 (2006): 187–216, on 198–199, 211–212.

6. See the discussion of the burning glass owned by the Lyonnais François Villette and his sons in Royal Society (Great Britain), *The Philosophical Transactions and Collections, to the end of the year 1720,* 4th ed., 5 vols. (London: 1732), 1:212.

7. On Duval, see R. Limouzin-Lamothe, "Duval, Jean Baptiste," in *Dictionnaire de Biographie Française,* ed. J. Balteau, M. Barroux, and M. Prévost, 19 vols. to date (Paris: Letouzey et Ané, 1933–), 12: cols. 973–975. For his stay in Italy, see F. G. Pariset, "Les remarques triennales de J.-B. du Val," *Revue de la Méditerranée* 15, no. 3 (1955): 237–260; 15, no. 4 (1955): 354–372; 15, no. 5 (1955): 481–499; 15, no. 6 (1955): 585–600; and "Jean-Baptiste du Val en Italie, 1608–1609," *Revue de la Méditerranée* 20, nos. 3–4 (1960): 215–227; 21 (1961): 23–40, 171–192, 255–267.

8. Pariset, "Jean Baptiste du Val en Italie: De Venise à Florence," *Revue de la Méditerranée* 20, nos. 3–4 (1960): 226–227.

9. *Carteggio inedito di Ticone Brahe, Giovanni Keplero e di altri celebri astronomi e matematici dei secoli XVI e XVII con Giovanni Antonio Magini,* ed. Antonio Favaro (Bologna: Nicola Zanichelli, 1886), 448–449. All translations are mine, unless otherwise indicated.

10. For the letters presenting the telescope to Henri IV and to the Duke de Sully, see Albert van Helden, *The Invention of the Telescope* (Philadelphia: American Philosophical Society, 1977), 43.

11. Henri Estienne, [*Francofordiense Emporium, sive Francofordienses Nundinae*] *La Foire de Francfort,* trans. Isidore Liseux (Paris: Isidore Liseux, 1875), 60–61. On the general character of this fair in the early modern period, see Harrison Clifford Dale, "The Frankfurt Book Fair," *Washington University Studies* 1 (1913): 54–65.

12. *Carteggio inedito,* 449–450. Giulio Mazarini SJ was the uncle of Mazarin.

13. *Carteggio inedito,* 251.

14. For a translation and discussion of the passage from Marius's *Mundus iovialis,* see Van Helden, *Invention of the Telescope,* 21, 47–48.

15. On these patent-seekers, see Van Helden, *Invention of the Telescope,* 20–25.

16. On Gastone Spinola, Count of Bruay, a collector of mathematical instruments, see Nicolas Claude Fabri de Peiresc, *Correspondence,* ed. Tamizey de Larroque, 7 vols. (Paris: Imprimerie Nationale, 1888–1898), 6:685, 692. Gastone Spinola's mathematical expertise is also mentioned in the Latin text accompanying the map of the Duchy of Limburg in the 1612 edition of Abraham Ortelius's *Theatrum orbis terrarum.* On the Spinola clan, see A. Miraeus, *Gentis Spinulae illustrium elogia* (Cologne: Johannes Kinckius, 1611), esp. 17–18, 34–36.

17. See Engel Sluiter, "The Telescope before Galileo," *Journal for the History of Astronomy* 28 (1997): 223–234, on 226.

18. *Carteggio inedito,* 448.

19. Galileo, *Opere,* 10:345.

20. Ibid., 342–343.

21. Ibid., 379, 384, 391–392, 418.

22. Ibid., 450.

23. Magini, *Briefve Instruction,* v. This version is also in Ausonio's *Theoretical Discourse on the Concave Spherical Mirror.*

24. Galileo, *Opere,* 10:357, 381; 13:49–50, 56, 263, 264, 266, 270; and Piero E. Ariotti, "Bonaventura Cavalieri, Marin Mersenne, and the Reflecting Telescope," *Isis* 66 (1975): 303–321, on 315–316.

25. Martin Horky, *Brevissima peregrinatio contra Nuncium Sidereum,* in Galileo, *Opere,* 3:(1):127–145, on 136.

26. Horky, *Brevissima peregrinatio,* in Galileo, *Opere,* 3:(1):144. For the

reaction of an Englishman in Padova to Horky's *Foray,* see Antonio Favaro, "Un inglese a Padova al tempo di Galileo: Lo 'Sloane Mss. 682' del British Museum," *Atti e memorie della reale academia di scienze, lettere ed arti in Padova,* n.s. (1917–1918): 34, 10–12, on 11. See also the Scotsman John Wedderburn's response in Galileo, *Opere,* 3(1):170–171.

27. On Callimachus's role at the Library at Alexandria, see Daniel Heller-Roazen, "Tradition's Destruction: On the Library of Alexandria," *October* 100 (2002): 133–153, on 143–145. On paradoxographia, see Alessandro Giannini, "Da Callimaco all'età imperiale: la letteratura paradossografica," *Acme* 17, no. 1 (1964): 99–140; and Emilio Gabba, "True History and False History in Classical Antiquity," *Journal of Roman Studies* 71 (1981): 50–62.

28. It is also possible that Horky knew from Magini of the unrealized plan of the Venetian Council of Ten to send an astronomer to Egypt to make observations for Tycho Brahe; on this venture, see Isabelle Pantin, "New Philosophy and Old Prejudices: Aspects of the Reception of Copernicanism in a Divided Europe," *Studies in the History and Philosophy of Science* 30, no. 2 (1999): 237–262, on 245–246.

29. Horky, *Peregrinatio,* in Galileo, *Opere,* 3(1):138.

30. Francesco Sizi, *Dianoia astronomica, optica, physica,* in Galileo, *Opere,* 3(1):201–250, on 238.

31. Prospero Alpini, *Historiae Aegypti naturali,* 2 vols. (Leiden: Gerard Potvliet, 1735), 1:4.

32. Ibid., 1:88. On this point, see Sonja Brentjes, "The Interest of the Republic of Letters in the Middle East, 1550–1700," *Science in Context* 12, no. 3 (1999): 435–468, on 446.

33. An exception is Edme-François Jomard et al., *Description de l'Egypte ou Recueil des observations et des recherches qui ont été faites en Egypte pendant l'expédition de l'armée française,* 2nd ed. (Paris:

C. L. F. Panckoucke, 1829), 9:257; there, under the heading "Lu-nettes à longue vue, telescopes," it is stated that "Ptolemy had a glass [*une lunette*] with which he saw ships at sea from an immense distance, and it was infinitely more valuable than the [sighting] tubes used before him. The discovery of the properties of glass [lenses] of different shapes may have made Jacob Metius and Galileo, who lived in Holland and in Italy, suspect around the year 1609 that these lenses had figured in the composition of Ptolemy's glass. Each of them, without consulting the other, arranged a convex objective lens and a concave eyepiece at a suitable distance in a tube, and the result was the telescope, and soon afterward, these [aerial or tubeless] telescopes which [Christiaan] Huygens and several other astronomers have perfected, and which, certainly, are not inferior to that of Ptolemy."

34. Cesi wrote to Francesco Stelluti, like Galileo and della Porta a member of the Academy of Lynxes, that he had arranged rooms in his Roman residence "for the studies and sciences which I am now undertaking, and above all those involving della Porta and mirrors." See Giuseppe Gabrieli, "Il Carteggio Linceo: Parte II (Anni 1610–1624)," *Atti della Reale Accademia Nazionale dei Lincei, Memorie della classe di scienze morali, storiche e filologiche,* ser. 6, 7 (1938): 123–522, on 160.

35. Johannes Henrici Alsted, *Methodus admirandorum mathematicorum* (Herborn in Nassau: n.p., 1613), 429–430.

36. Ben Jonson, *Staple of News,* 3.2; and Nunzio Vaccaluzzo, *Galileo Galilei nella poesia del suo secolo* (Milan: Remo Sandron, 1910), lx–lxiii.

37. *Iusta funebri Ptolomaei Oxoniensis T. Bodleii equities celebrati in Academiâ Oxoniensis mensis Martii 29 1613* (Oxford: J. Barnes, 1613), 115. I thank Robert Goulding for alerting me to this poem. On the establishment of the Bodleian, see R. A. Beddard, "The Official In-

auguration of the Bodleian Library on 8 November 1602," *Library* 3, no. 3 (2002): 255–283.

38. See Mordechai Feingold, *The Mathematicians' Apprenticeship: Science, Universities and Society in England 1560–1640* (Cambridge: Cambridge University Press, 1984), 157–158; and Thomas Birch, *The History of the Royal Society*, 4 vols. (London: A. Millar, 1756–1757), 4:156. On the presence of English students in Padua, see Jonathan Woolfson, *Padua and the Tudors: English Students in Italy, 1485–1603* (Toronto: University of Toronto Press, 1998).

39. Galileo Galilei, *Dialogue concerning the Two Chief World Systems*, trans. Stillman Drake (Berkeley: University of California Press, 1967), 93.

Acknowledgments

$\mathcal{T}$HE acknowledgments that accompany a book whose focus is appropriation and misinterpretation are in some fashion scripted, for it could be imagined that I would merely need to list the names of individuals from whom I have received an archival find, an argument, a diagram, or opportunity and encouragement, and to follow up with the ritual insistence that the errors and distortions that result from such generosity are wholly my own. Though these two steps seem to me correct, their formulaic nature masks the scope of my gratitude to those who have helped me finish this work.

I would like to thank first of all Noel Swerdlow, for his laconic suggestion some years back that I "stick with Galileo," for his interest in this new narrative, and for his patient last-minute advice about the diagrams that accompany the work. I am also very grateful to a number of early modernists who have made it possible for me to "stick with Galileo," especially Raz Chen, Rivka Feldhay, Paula Findlen, David Freedberg, Michel Hochmann, Sachiko Kusukawa, Pamela O. Long, A. Mark Smith, Pamela Smith, Lucia Tomasi Tongiorgi, and Alessandro Tosi. Their queries, suggestions, corrections, and accommodation—

both literal and figurative—have greatly lowered the boundaries between their disciplines and my own, Comparative Literature, without, I hope, any erosion of the standards that govern all of our endeavors.

I owe special thanks to those who have particularly indulged me in my efforts to use older texts about legendary and real optical devices to explain the gaps in Galileo's own narrative. I thank Robert Goulding for sharing his expertise in early modern optics, for allowing me to discuss the myth of the Pharos of Alexandria within this context, and for realizing that a bad poem about "Galileo's mirror" would be just the sort of thing that would interest me. I am also grateful to three fellow travelers in the Galilean orbit for their intellectual generosity: Michele Camerota for his judicious, tactful, and helpful comments upon an earlier version of my work; Mario Biagioli for his general interest in alternative narratives about Galileo and for his special gift of composing titles; and Nick Wilding for his great expertise in the world of Giovanni Francesco Sagredo. I am especially grateful to Sven Dupré, whose foundational work on Ettore Ausonio has revised, and will continue to revise, the conventional views of Galileo's early acquaintance with optics, whose interest in the "Elizabethan telescope" has been so crucial to my own project, and whose great intellectual range as a historian of science is complemented by his kindness and open-mindedness.

I have benefited most of all, finally, from the intellectual guidance and constant friendship of Albert van Helden, whose own work on the invention of the Dutch telescope is the basis of and inspiration for my own, and who has cheerfully endured delays in our study of the sunspots in order to allow me to finish this project, and withstood and responded to the onslaught of email queries I have generated over the years. I consider myself extraordinarily lucky to have a colleague so willing to elucidate the obscure history of early modern optical instruments, the mysteries of academia, and the arcana of contemporary American politics in the course of a typical electronic message.

A number of individuals have gone to great lengths to help me get this manuscript into printed form. I thank first of all my editor at Harvard University Press, Michael Fisher, for the interest he took in the project, for his sound judgment about its revision, and for the remarkable efficiency and good cheer he has shown throughout this process. I am also very grateful to Anne Zarrella and Susan Abel for their help in guiding the work through the Press, and for their prompt attention to all my questions in this phase. And I especially appreciate the efforts of Margaret Nelson, who has managed to convert my very feeble sketches into two excellent illustrations, and the vigilance of Wendy Nelson, who in the course of a few short weeks has done a splendid job of correcting and improving my prose and my footnotes. Last, I am very indebted to Martha Mayou for compiling a splendid index in such a short order.

Finally, I would like to thank my children, Jimmy and John English, for their involvement with this book. Lest they (or others) imagine that their contribution was confined to their intermittent ability to let me work in peace, I remind them of their willingness to visit any library with "dangerous" compact shelving, and of their enthusiastic conversion of a large cardboard box into a *camera obscura* for a little optical experimentation. And to my husband, Jim English, for his shrewd contributions to an argument remote from his own field of British postmodernism, and for his inimitable combination of kindness and wit, I offer my greatest thanks. This book is dedicated to him, but as promised above, I claim the errors and distortions that remain as my own.

Index